ASE Test Preparation Series

Automobile Test

Electrical/Electronics Systems (Test A6)

4th Edition

DELMAR
CENGAGE Learning™

Australia   Canada   Mexico   Singapore   Spain   United Kingdom   United States

DELMAR
CENGAGE Learning™

ASE Test Preparation Series: Automobile Test: Electrical/Electronics Systems (Test A6), Fourth Edition

Vice President, Technology Professional Business Unit: Gregory L. Clayton

Product Development Manager: Kristen Davis

Product Manager: Kim Blakey

Editorial Assistant: Vanessa Carlson

Director ofMarketing: Beth A. Lutz

Marketing Specialist: Brian McGrath

Marketing Coordinator: Marissa Maiella

Production Manager: Andrew Crouth

Production Editor: Kara A. DiCaterino

Senior Project Editor: Christopher Chien

XML Architect: Jean Kaplansky

Cover Design: Michael Egan

Cover Images:P ortion courtesy of DaimlerChryslerCorporation

For product information and technology assistance, contact us at
Cengage Learning Customer & Sales Support, 1-800-354-9706

For permission to use material from this text or product, submit all requests online at **www.cengage.com/permissions**
Further permissions questions can be emailed to **permissionrequest@cengage.com**

ISBN-13: 978-1-4180-3883-0

ISBN-10: 1-4180-3883-0

Delmar
Executive Woods
5 Maxwell Drive
Clifton Park, NY 12065
USA

Cengage Learning is a leading provider of customized learning solutions with office locations around the globe, including Singapore, the United Kingdom, Australia, Mexico, Brazil, and Japan. Locate your local office at **international.cengage.com/region**

Cengage Learning products are represented in Canada by Nelson Education, Ltd.

For your course and learning solutions, visit **delmar.cengage.com**

Visit our corporate website at **www.cengage.com**

Notice to the Reader

Publisher does not warrant or guarantee any of the products described herein or perform any independent analysis in connection with any of the product information contained herein. Publisher does not assume, and expressly disclaims, any obligation to obtain and include information other than that provided to it by the manufacturer. The reader is expressly warned to consider and adopt all safety precautions that might be indicated by the activities described herein and to avoid all potential hazards. By following the instructions contained herein, the reader willingly assumes all risks in connection with such instructions. The publisher makes no representations or warranties of any kind, including but not limited to, the warranties of fitness for particular purpose or merchantability, nor are any such representations implied with respect to the material set forth herein, and the publisher takes no responsibility with respect to such material. The publisher shall not be liable for any special, consequential, or exemplary damages resulting, in whole or part, from the readers' use of, or reliance upon, this material.

Printed in the United States of America
6 7 11 10

Contents

Section 1   The History and Purpose of ASE

Section 2   Take and Pass Every ASE Test

Section 3   Types of Questions on an ASE Exam

Section 4   Overview of the Task List

Section 5 Sample Test for Practice

Section 6 Additional Test Questions for Practice

Section 7 Appendices

Preface

Delmar Cengage Learning is very pleased that you have chosen our ASE Test Preparation Series to prepare yourself for the automotive ASE Examination. These guides are available for all of the automotive areas including A1–A8, the L1 Advanced Diagnostic Certification, the P2 Parts Specialist, the C1 Service Consultant and the X1 Undercar Specialist. These guides are designed to introduce you to the Task List for the test you are preparing to take, give you an understanding of what you are expected to be able to do in each task, and take you through sample test questions formatted in the same way the ASE tests are structured.

If you have a basic working knowledge of the discipline you are testing for, you will find Delmar's ASE Test Preparation Series to be an excellent way to understand the "must know" items to pass the test. These books are not textbooks. Their objective is to prepare the technician who has the requisite experience and schooling to challenge ASE testing. It cannot replace the hands-on experience or the theoretical knowledge required by ASE to master vehicle repair technology. If you are unable to understand more than a few of the questions and their explanations in this book, it could be that you require either more shop-floor experience or further study. Some resources that can assist you with further study are listed on the rear cover of this book.

Each book begins with an item-by-item overview of the ASE Task List with explanations of the minimum knowledge you must possess to answer questions related to the task. Following that there are 2 sets of sample questions followed by an answer key to each test and an explanation of the answers to each question. A few of the questions are not strictly ASE format but were included because they help teach a critical concept that will appear on the test. We suggest that you read the complete Task List Overview before taking the first sample test. After taking the first test, score yourself and read the explanation to any questions that you were not sure about, including the questions you answered correctly. Each test question has a reference back to the related task or tasks that it covers. This will help you to go back and read over any area of the task list that you are having trouble with. Once you are satisfied that you have all of your questions answered from the first sample test, take the additional tests and check them. If you pass these tests, you will be prepared to do well on the ASE test.

Our Commitment to Excellence

The 4th edition of Delmar's ASE Test Preparation Series has been through a major revision with extensive updates to the ASE's task lists, test questions, and answers and explanations. Delmar Cengage Learning has sought out the best technicians in the country to help with the updating and revision of each of the books in the series.

About the Series Advisor

To promote consistency throughout the series, a series advisor took on the task of reading, editing, and helping each of our experts give each book the highest level of accuracy possible. Dan Perrin has served in the role of Series Advisor for the 4th edition of the ASE Test Preparation Series. Dan began ASE testing with the first series of tests in 1972 and has been continually certified ever since. He holds ASE master status in automotive, truck, collision, and machinist. He is also L1, L2, and alternated fuels certified, along with some others that have expired. He has been an automotive educator since 1979, having taught at the secondary, post-secondary, and industry levels. His service includes participation on boards that include the North American Council of Automotive Teachers (NACAT), the Automotive Industry Planning Council (AIPC), and the National Automotive Technicians Education Foundation (NATEF). Dan currently serves as the Executive Manager of NACAT and Director of the NACAT Education Foundation.

Thanks for choosing Delmar's ASE Test Preparation Series. All of the writers, editors, Delmar Staff, and myself have worked very hard to make this series second to none. I know you are going to find this book accurate and easy to work with. It is our objective to constantly improve our product at Delmar by responding to feedback.

If you have any questions concerning the books in this series, you can email me at: autoexpert@trainingbay.com.

Dan Perrin
Series Advisor

1

The History and Purpose of ASE

ASE began as the National Institute for Automotive Service Excellence (NIASE). It was founded as a non-profit independent entity in 1972 by a group of industry leaders with the single goal of providing a means for consumers to distinguish between incompetent and competent technicians. It accomplishes this goal by testing and certification of repair and service professionals. From this beginning it has evolved to be known simply as ASE (Automotive Service Excellence) and today offers more than 40 certification exams in automotive, medium/heavy duty truck, collision, engine machinist, school bus, parts specialist, automobile service consultant, and other industry-related areas. At this time there are more than 400,000 professionals with current ASE certifications. These professionals are employed by new car and truck dealerships, independent garages, fleets, service stations, franchised service facilities, and more. ASE continues its mission by also providing information that helps consumers identify repair facilities that employ certified professionals through its Blue Seal of Excellence Recognition Program. Shops that have a minimum of 75% of their repair technicians ASE certified and meet other criteria can apply for and receive the Blue Seal of Excellence Recognition from ASE.

ASE recognized that educational programs serving the service and repair industry also needed a way to be recognized as having the faculty, facilities, and equipment to provide a quality education to students wanting to become service professionals. Through the combined efforts of ASE, industry, and education leaders, the non-profit National Automotive Technicians Education Foundation (NATEF) was created to evaluate and recognize training programs. Today more than 2000 programs are ASE certified under the standards set by the service industry. ASE/NATEF also has a certification of industry (factory) training program known as CASE. CASE stands for Continuing Automotive Service Education and recognizes training provided by replacement parts manufacturers as well as vehicle manufacturers.

ASE certification testing is administered by the American College Testing (ACT). Strict standards of security and supervision at the test centers insure that the technician who holds the certification earned it. Additionally ASE certification also requires that the person passing the test to be able to demonstrate that they have two years of work experience in the field before they can be certified. Test questions are developed by industry experts that are actually working in the field being tested. There is more detail on how the test is developed and administered in the next section. Paper and pencil tests are administered twice a year at over seven hundred locations in the United States. Computer based testing is now also available with the benefit of instant test results at certain established test centers. The certification is valid for five years and can be recertified by retesting. So that consumers can recognize certified technicians, ASE issues a jacket patch, certificate, and wallet card to certified technicians and makes signs available to facilities that employ ASE certified technicians.

You can contact ASE at any of the following:

National Institute for Automotive Service Excellence
101 Blue Seal Drive S.E.
Suite 101
Leesburg, VA 20175
Telephone 703-669-6600
FAX 703-669-6123
www.ase.com

2 | Take and Pass Every ASE Test

Participating in an Automotive Service Excellence (ASE) voluntary certification program gives you a chance to show your customers that you have the "know-how" needed to work on today's modern vehicles. The ASE certification tests allow you to compare your skills and knowledge to the automotive service industry's standards for each specialty area.

If you are the "average" automotive technician taking this test, you are in your mid-thirties and have not attended school for about fifteen years. That means you probably have not taken a test in many years. Some of you, on the other hand, have attended college or taken postsecondary education courses and may be more familiar with taking tests and with test-taking strategies. There is, however, a difference in the ASE test you are preparing to take and the educational tests you may be accustomed to.

How are the tests administered?

ASE test are administered at over 750 test sites in local communities. Paper and pencil tests are the type most widely available to technicians. Each tester is given a booklet containing questions with charts and diagrams where required. You can mark in this test booklet but no information entered in the booklet is scored. Answers are recorded on a separate answer sheet. You will enter your answers, using a number 2 pencil only. ASE recommends you bring four sharpened number 2 pencils that have erasers. Answer choices are recorded by coloring in the blocks on the answer sheet. The answer sheets are scanned electronically and the answers tabulated. For test security, test booklets include randomly generated questions. Your answer key must be matched to the proper booklet so it is important to correctly enter the booklet serial number on the answer sheet. All instructions are printed on the test materials and should be followed carefully.

ASE has introduced Computer Based Testing (CBT) at some locations. While the test content is the same for both testing methods the CBT tests have some unique requirements and advantages. It is strongly recommended that technicians considering the CBT tests go the ASE web page at www.ASE.com and review the conditions and requirements for this type of test. There is a demonstration of a CBT that allows you to experience this type of test before you register. Some technicians find this style of testing provides an advantage, while others find operating the computer a distraction. One significant benefit of CBT is the availability of instant results. You can receive your test results before you leave the test center. CBT testing also offers increased flexibility in scheduling. The cost for taking CBTs is slightly higher than paper and pencil tests and the number of testing sites is limited. The first time test taker may be more comfortable with the paper and pencil tests but technicians now have a choice.

Who Writes the Questions?

The questions are written by service industry experts in the area being tested. Each area will have its own technical experts. Questions are entirely job related. They are designed to test the skills you need to be a successful technician. Theoretical knowledge is important and necessary to answer the questions, but the ability to apply that knowledge is the basis of ASE test questions.

Each question has its roots in an ASE "item-writing" workshop where service representatives from automobile manufacturers (domestic and import), aftermarket parts and equipment manufacturers, working technicians, and vocational educators meet in a workshop setting to share ideas and translate them into test questions. Each test question written by these experts must survive review by all members of the group.

The questions are written to deal with practical application of soft skills and system knowledge experienced by technicians in their day-to-day work.

All questions are pre-tested and quality-checked on a national sample of technicians. Those questions that meet ASE standards of quality and accuracy are included in the scored sections of the tests; the "rejects" are sent back to the drawing board or discarded altogether.

Each certification test is made up of between forty and eighty multiple-choice questions.

Note: Each test could contain additional questions that are included for statistical research purposes only. Your answers to these questions will not affect your score, but since you do not know which ones they are, you should answer all questions on the test. The five-year Recertification Test will cover the same content areas as those listed above. However, the number of questions in each content area of the Recertification Test will be reduced by about one-half.

Objective Tests

A test is called an objective test if the same standards and conditions apply to everyone taking the test and there is only one correct answer to each question.

Objective tests primarily measure your ability to recall information. A well-designed objective test can also test your ability to understand, analyze, interpret, and apply your knowledge. Objective tests include true-false, multiple choice, fill in the blank, and matching questions. ASE's tests consist exclusively of four-part multiple-choice objective questions.

The following are some strategies that may be applied to your tests.

Before beginning to take an objective test, quickly look over the test to determine the number of questions, but do not try to read through all of the questions. In an ASE test, there are usually between forty and eighty questions, depending on the subject. Read through each question before marking your answer. Answer the questions in the order they appear on the test. Leave the questions blank that you are not sure of and move on to the next question. You can return to those unanswered questions after you have finished the others. They may be easier to answer at a later time after your mind has had additional time to consider them on a subconscious level. In addition, you might find information in other questions that will help you recall the answers to some of them.

Do not be obsessed by the apparent pattern of responses. For example, do not be influenced by a pattern like **D, C, B, A, D, C, B, A** on an ASE test.

There is also a lot of folk wisdom about taking objective tests. For example, there are those who would advise you to avoid response options that use certain words such as *all, none, always, never, must,* and *only,* to name a few. This, they claim, is because nothing in life is exclusive. They would advise you to choose response options that use words that allow for some exception, such as *sometimes, frequently, rarely, often, usually, seldom,* and *normally.* They would also advise you to avoid the first and last option (A and D) because test writers, they feel, are more comfortable if they put the correct answer in the middle (B and C) of the choices. Another recommendation often offered is to select the option that is either shorter or longer than the other three choices because it is more likely to be correct. Some would advise you to never change an answer since your first intuition is usually correct.

Although there may be a grain of truth in this folk wisdom, ASE test writers try to avoid them and so should you. There are just as many **A** answers as there are **B** answers, just as many **D** answers as **C** answers. As a matter of fact, ASE tries to balance the answers at about 25 percent per choice **A, B, C,** and **D.** There is no intention to use "tricky" words, such as outlined above. Put no credence in the opposing words "sometimes" and "never," for example.

Multiple-choice tests are sometimes challenging because there are often several choices that may seem possible, and it may be difficult to decide on the correct choice. The best strategy, in this case, is to first determine the correct answer before looking at the options. If you see the answer you decided on, you should still examine the options to make sure that none seem more correct than yours. If you do not know or are not sure of the answer, read each option very carefully and try to eliminate those options that you know to be wrong. That way, you can often arrive at the correct choice through a process of elimination.

If you have gone through all of the test and you still do not know the answer to some of the questions, <u>then guess.</u> Yes, guess. You then have at least a 25 percent chance of being correct. If you leave the question blank, you have no chance. Your score is based on the number of questions answered correctly.

Preparing for the Exam

The main reason we have included so many sample and practice questions in this guide is, simply, to help you learn what you know and what you don't know. We recommend that you work your way through each question in this book. Before doing this, carefully look through Section 3; it contains a description and explanation of the question types you'll find on an ASE exam.

Once you understand what the questions will look like, move to the sample test. Answer one of the sample questions (Section 5) then read the explanation (Section 7) to the answer for that question. If you don't feel you understand the reasoning for the correct answer, go back and read the overview (Section 4) for the task that is related to that question. If you still don't feel you have a solid understanding of the material, identify a good source of information on the topic, such as a textbook, and do some more studying.

After you have completed all of the sample test items and reviewed your answers, move to the additional questions (Section 6). This time answer the questions as if you were taking an actual test. Do not use any reference or allow any interruptions in order to get a feel for how you will do on an actual test. Once you have answered all of the questions, grade your results using the answer key in Section 7. For every question that you gave a wrong answer to, study the explanations to the answers and/or the overview of the related task areas. Try to determine the root cause for your missing the question. The easiest thing to correct is learning the correct technical content. The hardest thing to correct is behaviors that lead you to a wrong conclusion. If you knew the information but still got it wrong there is a behavior problem that will need to be corrected. An example would be reading too quickly and skipping over words that affect your reasoning. If you can identify what you did that caused you to answer the question incorrectly you can eliminate that cause and improve your score. Here are some basic guidelines to follow while preparing for the exam:

- Focus your studies on those areas you are weak in.
- Be honest with yourself while determining if you understand something.
- Study often but in short periods of time.
- Remove yourself from all distractions while studying.
- Keep in mind the goal of studying is not just to pass the exam, the real goal is to learn!
- Prepare physically by getting a good night's rest before the test and eat meals that provide energy but do not cause discomfort.
- Arrive early to the test site to avoid long waits as test candidates check in and to allow all of the time available for your tests.

During the Test

On paper and pencil tests you will be placing your answers on a sheet where you will be required to color in your answer choice. Stray marks or incomplete erasures may be picked up as an answer by the electronic reader, so be sure only your answers end up on the sheet. One of the biggest problems an adult faces in test taking, it seems, is placing the answer in the correct spot on the answer sheet. Make certain that you mark your answer for, say, question 21, in the space on the answer sheet designated for the answer for question 21. A correct response in the wrong line will probably result in two questions being marked wrong, one with two answers (which could include a correct answer but will be scored wrong) and the other with no answer. Remember, the answer sheet on the written test is machine scored and can only "read" what you have colored in.

If you finish answering all of the questions on a test and have remaining time, go back and review the answers to those questions that you were not sure of. You can often catch careless errors by using the remaining time to review your answers. Carefully check your answer sheet for blank answer blocks or missing information.

At practically every test, some technicians will invariably finish ahead of time and turn their papers in long before the final call. Some technicians may be doing recertification tests and others may be taking fewer tests than you. Do not let them distract or intimidate you.

It is not wise to use less than the total amount of time that you are allotted for a test. If there are any doubts, take the time for review. Any product can usually be made better with some additional effort. A test is no exception. It is not necessary to turn in your test paper until you are told to do so.

Testing Time Length

An ASE written test session is four hours. You may attempt from one to a maximum of four tests in one session. It is recommended, however, that no more than a total of 225 questions be attempted at any test session. This will allow for just over one minute for each question.

Visitors are not permitted at any time. If you wish to leave the test room, for any reason, you must first ask permission. If you finish your test early and wish to leave, you are permitted to do so only during specified dismissal periods.

You should monitor your progress and set an arbitrary limit to how much time you will need for each question. This should be based on the number of questions you are attempting. It is suggested that you wear a watch because some facilities may not have a clock visible to all areas of the room.

Computer-Based Tests are allotted a testing time according to the number of questions ranging from one half hour to one and one half hours. Advanced level tests are allowed two hours. This time is by appointment and you should be sure to be on time to insure that you have all of the time allocated. If you arrive late for a CBT test appointment you will only have the amount of time remaining on your appointment.

Your Test Results!

You can gain a better perspective about tests if you know and understand how they are scored. ASE's tests are scored by American College Testing (ACT), a nonpartial, unbiased organization having no vested interest in ASE or in the automotive industry.

Each question carries the same weight as any other question. For example, if there are fifty questions, each is worth 2 percent of the total score. The passing grade is 70 percent. That means you must correctly answer thirty-five of the fifty questions to pass the test.

The test results can tell you:

- where your knowledge equals or exceeds that needed for competent performance, or

- where you might need more preparation.

Your ASE test score report is divided into content areas and will show the number of questions in each content area and how many of your answers were correct. These numbers provide information about your performance in each area of the test. However, because there may be a different number of questions in each content area of the test, a high percentage of correct answers in an area with few questions may not offset a low percentage in an area with many questions.

It should be noted that one does not "fail" an ASE test. The technician who does not pass is simply told "More Preparation Needed." Though large differences in percentages may indicate problem areas, it is important to consider how many questions were asked in each area. Since each test evaluates all phases of the work involved in a service specialty, you should be prepared in each area. A low score in one area could keep you from passing an entire test.

There is no such thing as average. You cannot determine your overall test score by adding the percentages given for each task area and dividing by the number of areas. It doesn't work that way because there generally are not the same number of questions in each task area. A task area with twenty questions, for example, counts more toward your total score than a task area with ten questions.

Your test report should give you a good picture of your results and a better understanding of your strengths and weaknesses for each task area.

If you fail to pass the test, you may take it again at any time it is scheduled to be administered. You are the only one who will receive your test score. Test scores will not be given over the telephone by ASE nor will they be released to anyone without your written permission.

3 | Types of Questions on an ASE Exam

ASE certification tests are often thought of as being tricky. They may seem to be tricky if you do not completely understand what is being asked. The following examples will help you recognize certain types of ASE questions and avoid common errors.

Paper-and-pencil tests and computer-based test questions are identical in content and difficulty. Most initial certification tests are made up of forty to eighty multiple-choice questions. Multiple-choice questions are an efficient way to test knowledge. To answer them correctly, you must think about each choice as a possibility, and then choose the one that best answers the question. To do this, read each word of the question carefully. Do not assume you know what the question is about until you have finished reading it.

About 10 percent of the questions on an actual ASE exam will use an illustration. These drawings contain the information needed to correctly answer the question. The illustration must be studied carefully before attempting to answer the question. Often, technicians look at the possible answers then try to match up the answers with the drawing. Always do the opposite; match the drawing to the answers. When the illustration is showing an electrical schematic or another system in detail, look over the system and try to figure out how the system works before you look at the question and the possible answers.

Multiple-Choice Questions

The most common type of question used on ASE Tests is the multiple-choice question. This type of question contains three "distracters" (wrong answers) and one "key" (correct answer). When the questions are written effort is made to make the distracters plausible to draw an inexperienced technician to one of them. This type of question gives a clear indication of the technician's knowledge. Using multiple criteria including cross-sections by age, race, and other background information, ASE is able to guarantee that a question does not bias for or against any particular group. A question that shows bias toward any particular group is discarded. If you encounter a question that you are unsure of, reverse engineer it by eliminating the items that it cannot be. For example:

A rocker panel is a structural member of which vehicle construction type?

A. Front-wheel drive
B. Pickup truck
C. Unibody
D. Full-frame

Analysis:

This question asks for a specific answer. By carefully reading the question, you will find that it asks for a construction type that uses the rocker panel as a structural part of the vehicle.

Answer A is wrong. Front-wheel drive is not a vehicle construction type.
Answer B is wrong. A pickup truck is not a type of vehicle construction.
Answer C is correct. Unibody design creates structural integrity

by welding parts together, such as the rocker panels, but does not require exterior cosmetic panels installed for full strength.

Answer D is wrong. Full-frame describes a body-over-frame construction type that relies on the frame assembly for structural integrity.

Therefore, the correct answer is C. If the question was read quickly and the words "construction type" were passed over, answer A may have been selected.

EXCEPT Questions

Another type of question used on ASE tests has answers that are all correct except one. The correct answer for this type of question is the answer that is wrong. The word "**EXCEPT**" will always be in capital letters. You must identify which of the choices is the wrong answer. If you read quickly through the question, you may overlook what the question is asking and answer the question with the first correct statement. This will make your answer wrong. An example of this type of question and the analysis is as follows:

All of the following are tools for the analysis of structural damage **EXCEPT:**

A. height gauge
B. tape measure.
C. dial indicator.
D. tram gauge.

Analysis:

The question really requires you to identify the tool that is not used for analyzing structural damage. All tools given in the choices are used for analyzing structural damage except one. This question presents two basic problems for the test-taker who reads through the question too quickly. It may be possible to read over the word "**EXCEPT**" in the question or not think about which type of damage analysis would use answer C. In either case, the correct answer may not be selected. To correctly answer this question, you should know what tools are used for the analysis of structural damage. If you cannot immediately recognize the incorrect tool, you should be able to identify it by analyzing the other choices.

Answer A is wrong. A height gauge may be used to analyze structural damage.
Answer B is wrong. A tape measure may be used to analyze structural damage.
Answer C is correct. A dial indicator may be used as a damage analysis tool for moving parts, such as wheels, wheel hubs, and axle shafts, but would not be used to measure structural damage.
Answer D is wrong. A tram gauge is used to measure structural damage.

Technician A, Technician B Questions

The type of question that is most popularly associated with an ASE test is the "Technician A says . . . Technician B says . . . Who is right?" type. In this type of question, you must identify the correct statement or statements. To answer this type of question correctly, you must carefully read each technician's statement and judge it on its own merit to determine if the statement is true.

Sometimes this type of question begins with a statement about some analysis or repair procedure. This is often referred to as the stem of the question and provides the setup or background information required to understand the conditions the question is based on. This is followed by two statements about the cause of the concern, proper inspection, identification, or repair choices. You are asked whether the first statement, the second statement, both statements, or neither statement is correct. Analyzing this type of question is a little easier than the other types because there are only two ideas to consider although there are still four choices for an answer.

Technician A, Technician B questions are really double true or false questions. The best way to analyze this kind of question is to consider each technician's statement separately. Ask yourself, is A true or false? Is B true or false? Then select your answer from the four choices. An important point to remember is that an ASE Technician A, Technician B question will never have Technician A and B directly disagreeing with each other. That is why you must evaluate each statement independently.

An example of this type of question and the analysis of it follows.

A vehicle comes into the shop with a gas gauge that will not register above one half full. When the sending unit circuit is disconnected the gauge reads empty and when it is connected to ground the gauge goes to full. Technician A says that the sending unit is shorted to ground. Technician B says the gauge circuit is working and the sending unit is likely the problem. Who is right?

A. A only
B. B only
C. Both A and B
D. Neither A nor B

Analysis:

Reading of the stem of the question sets the conditions of the customer concern and establishes what information is gained from testing. General knowledge of gauge circuits and test procedures are needed to correctly evaluate the technician's conclusions. Note: Avoid being distracted by experience with unusual or problem vehicles that you may have worked on, Other technicians taking the same test do not have that knowledge, so it should not be used as the basis of your answers.

Technician A is wrong because a shorted to ground sending unit would produce a gauge reading equivalent to the test conditions of a grounding the circuit and produce a full reading. **Technician B is correct** because the gauge spans when going from an open circuit to a completely
grounded circuit. This would tend to indicate that the problem had to be in the sending unit.
Answer C is not correct. Both technicians are identifying the problem as a sending unit but technician A qualified the problem as a specific type of failure (grounded) that would not have caused the symptoms of the vehicle.
Answer D is not correct because technician B's diagnosis is a possible cause of the conditions identified.

Most-Likely Questions

Most-Likely questions are somewhat difficult because only one choice is correct while the other three choices are nearly correct. An example of a Most-Likely-cause question is as follows:

The Most-Likely cause of reduced turbocharger boost pressure may be a:

A. wastegate valve stuck closed.
B. wastegate valve stuck open.
C. leaking wastegate diaphragm.
D. disconnected wastegate linkage.

Analysis:

Answer A is wrong. A wastegate valve stuck closed increases turbocharger boost pressure.
Answer B is correct. A wastegate valve stuck open decreases turbocharger boost pressure.
Answer C is wrong. A leaking wastegate valve diaphragm increases turbocharger boost pressure.

Answer D is wrong. A disconnected wastegate valve linkage will increase turbocharger boost pressure.

LEAST-Likely Questions

Notice that in Most-Likely questions there is no capitalization. This is not so with LEAST-Likely type questions. For this type of question, look for the choice that would be the LEAST-Likely cause of the described situation. Read the entire question carefully before choosing your answer. An example is as follows:

What is the LEAST-Likely cause of a bent pushrod?

A. Excessive engine speed
B. A sticking valve
C. Excessive valve guide clearance
D. A worn rocker arm stud

Analysis:

Answer A is wrong. Excessive engine speed may cause a bent pushrod.
Answer B is wrong. A sticking valve may cause a bent pushrod.
Answer C is correct. Excessive valve clearance will not generally cause a bent pushrod.
Answer D is wrong. A worn rocker arm stud may cause a bent pushrod.

You should avoid relating questions to those unusual situations that you may have encountered and answer based on the technical and mechanical possibilities.

Summary

There are no four-part multiple-choice ASE questions having "none of the above" or "all of the above" choices. ASE does not use other types of questions, such as fill-in-the-blank, completion, true-false, word-matching, or essay. ASE does not require you to draw diagrams or sketches. If a formula or chart is required to answer a question, it is provided for you. There are no ASE questions that require you to use a pocket calculator.

Electrical/Electronic Systems (Test A6)

The following section includes the task areas and task lists for this test and a written overview of the topics covered in the test.

The task list describes the actual work you should be able to do as a technician that you will be tested on by the ASE. This is your key to the test, and you should review this section carefully. We have based our sample test and additional questions upon these tasks, and the overview section will also support your understanding of the task list. ASE advises that the questions on the test may not equal the number of tasks listed; the task lists tell you what ASE expects you to know how to do and be ready to be tested upon.

At the end of each question in the Sample Test and Additional Test Questions sections, a letter and number will be used as a reference back to this section for additional study. Note the following example: **E.1.8.**

E. Lighting Systems Diagnosis and Repair (6 Questions)

Task E.1   **Headlights, Parking Lights, Tail Lights, Dash Lights, and Courtesy Lights (3 Questions)**

Task E.1.8   **Inspect, test, and repair or replace switches, relays, bulbs, sockets, connectors, wires, controllers, and printed circuit boards of instrument lighting circuits.**

Example:
1. Which of the following are true about light circuits?
   A. Signal and stoplights often share a common bulb filament.
   B. Many back-up lights and stoplights often share the same bulb.
   C. The taillight and stoplights share a common bulb filament.
   D. When the brake pedal is depressed, the stoplight switch is open.   (E.1.8)

Analysis:

Question #1
Answer A is correct. This question contains a summary of all of the bulb circuits mentioned in the question. Typically the turn signal and brake lights share a common bulb filament.
Answer B is wrong. Answer B is wrong because backup lamps normally have a single filament bulb in its own housing.
Answer C is wrong. Answer C suggests that the taillights and stoplight share a filament. This is not true; the brake lights operate brighter than the taillights and use a separate filament.
Answer D is wrong. Answer D is wrong because the brake pedal switch closes the circuit when the pedal is depressed.

Automotive Electrical—The Basics

All electrical circuits live in a world that has rules. While it may not always seem that way, there is an underlying rule to every electrical circuit, even when it is not working it is obeying those rules. It is based in a discipline called physics. As technicians we have to understand a fair amount of physics whether we realize it or not. One example of a law of physics is that if you do not have a fender cover on a car, the screwdriver will always slip out of your hand and scratch the fender. Seriously though there are some basics that, once you understand them, will make it possible to understand almost any automotive electrical circuit by simply looking at the wiring diagram. Due to the vast number of tasks in the ASE A6 Electrical Task List, we are going to start with those basics so that our overview of each task will have some common ground (all puns intended) that will make it possible for you to understand each task with a minimum of repetition. Let's get back to the law, and let's discuss the one law that no electrical circuit can break or work without; Ohm's law. For our purposes the history of Ohm's law is not important. What is important is what it teaches us. It reads like this:

"The amount of current flowing in a circuit made up of pure resistances is directly proportional to the electromotive forces impressed on the circuit and inversely proportional to the total resistance of the circuit."

Wait, don't panic yet. Let's make this easier to understand. Most of the complexity of this lies in everyday words that you already understand. Simply put, Ohm's law is the relationship between voltage, current, and amperage. They allow us to test things because, if we think of it as a triangle and we know two sides of the triangle, then we can test the 3rd side using Ohm's law. To show you how that is true, let's gain a better understanding of each side of the triangle.

Voltage

Voltage is the electrical force, or "pressure," that causes current to flow in a circuit. It is measured in "volts" (V or E). Using a water line as an analogy voltage would be the force that is pushing the water (electrons) through the line.

Current

Current is the movement of electrical charge (the flow of electrons through the electronic circuit. Current is measured in "amperes" (amps, A or I). Current would be the flow of water moving through the tube (wire).

Resistance

Resistance is anything that causes an opposition to the flow of electricity in a circuit. It is used to control the amount of voltage and/or amperage in a circuit. Everything in the circuit causes a resistance (even wire). It is measured in "Ohms" (). Resistance is the size of the water line. The larger the line is, the less resistance, and the smaller the line, the greater the resistance. Or to put it back into electrical terms, the larger the pipe, the more electrons we can get through it, and with a smaller pipe we can get fewer electrons through it.

Be sure that you understand this before you move on. When you really get this clear, electrical service gets pretty easy. Let's look at some diagrams of how Ohm's law works.

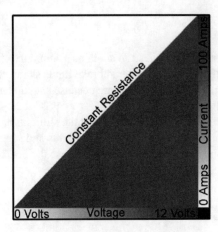

In this diagram we see that in a circuit that has constant resistance, a steady rise in voltage produces a steady rise in current. We'll call this the faucet theory: the more you turn on the water (voltage)

the faster it comes out (current), right up to the limits of the size of pipe (resistance) and how much pressure you can generate (voltage). Let's take the real-world example of a battery cable. Since we know that a car might take peaks of 250 amps to turn the engine, and we know that our battery is a 12-volt battery, we can understand that to get 250 amps worth of current with only 12 volts of pressure, we need a pretty big pipe. This is why battery cables are large on automobiles. If we were to double our voltage, we could get significantly more current through the same size pipe (battery cable).

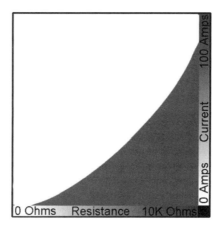

Since we know that our car has a 12-volt system (approximately 14.3 volts when the car is running), we take that as the amount of pressure we have to get any work we need done in our electrical system. Let's say we want to make a light bulb on the dash dimmer than it is when full battery voltage is applied to it. How do we do that? We add resistance to the circuit. Resistance causes the voltage to drop to the light bulb, and it gets dimmer, right to the point where we have added so much resistance that there is no voltage anymore. Now what happens to current all the while? As resistance is increased current flow decreases. Remember our water pipe analogy, if you add resistance (make the pipe smaller) there is less flow.

Now let's add one more rule to our set of rules for automotive electronics that is most evident in the charging system. When amperage goes up, voltage goes down. The opposite is true. Even generators must conform to Ohm's law. The harder a generator is working to create current to supply the vehicle or recharge the battery, the lower the voltage will go on the system. Let's go back to our faucet again. If you turn both handles on all the way and then go to another faucet and turn it on all the way, you will notice that the amount of water (current) coming out of each is less than when there was only one faucet on. That is because you have a set amount of water pressure (voltage), and the more you create demand for the water (current), the less pressure (voltage) you can maintain.

Study this section before you move on because this is the basis of all of the tasks in the ASE A6 Electrical test.

Task List and Overview

A. General Electrical/Electronic System Diagnosis (13 Questions)

Task A.1   **Check electrical circuits with a test light; determine needed repairs.**

Simple circuits can be tested using a test light. Test lights work well when testing continuity in higher amperage circuits that do not use any kind of solid-state or digital controls to switch them. Test lights, not surprisingly, work well to test light circuits, horn circuits, and for quickly checking to see if a power or ground circuit has the presence of electricity on it. The test light is simply connected to the opposite pole of the battery from the circuit you are testing. If you are testing the positive side of a light circuit, it will be connected to ground. If the test light comes on, you have a complete circuit. If the light is dim or does not come on, you can turn the circuit around and test the ground side by connecting the light to a good battery positive (B+) source. If the light is good there, then you know you have a good ground but a problem with the power side of your circuit. The limits of the test light are that you cannot really tell how much of a voltage drop you are experiencing based solely on how dim or bright the bulb. It is a handy, quick tool if you are familiar with the circuit you are testing, but it can become a source of damage or at least blown fuses if you happen to connect to a circuit that has the opposing side of the battery on it. To cut down on the potential for damage, it is really best to use a digital voltmeter to test because it provides you with a safer way to test and a way to test the other sides of Ohm's law by just rotating the switch.

Task A.2   Check voltages and voltage drops in electrical/electronic circuits; interpret readings and determine needed repairs.

A voltmeter may be connected across a component in a circuit to measure the voltage drop across the component. Current must be flowing through the circuit during the voltage drop test. The amount of voltage drop depends on the resistance in the component and the amount of current flow the component requires. Remember that if you are testing for a voltage drop on a circuit that is normally a complete path from B+ to ground, like a light, you will see an almost 100 percent voltage drop if the connections are good. Let us walk through a connection to perform a voltage drop test in the diagram.

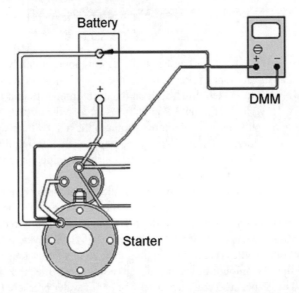

In the figure, the negative test lead for the digital multimeter (DMM) is connected to the negative post of the battery and the positive test lead of the DMM is connected to the ground lug of the starter. We are performing a voltage drop on the negative battery cable and the connection of the battery terminal to the negative battery post. Think of using the test leads as drawing brackets that surround the item(s) you want to test. Everything between the test leads is being tested. If everything is good, we do not want to see more than 0.2 volt loss in this critical connection. Now remember that you do not have a true reading of voltage drop until you exercise the circuit (it must be operating). What happens if we use our test leads to bracket a component that does work such as a light? If you are on both sides of a light bulb, or similar full power component, you will see almost a complete voltage drop to 0 volts, and that is normal and correct. Many ASE test questions will show you an open circuit where the DMM is reading 12 volts, and that is correct. They may also show you a circuit that is closed with a DMM reading of nearly 0 volts, and that is correct. Remember Ohm's law: more current means less voltage, and less current means more voltage.

Task A.3   Check current flow in electrical/electronic circuits and components; interpret readings and determine needed repairs.

An ammeter has low internal resistance so that it can measure current through it. The meter must be connected in series with a circuit. Some ammeters have an inductive clamp that fits over a wire in the circuit. These ammeters measure the current flow from the strength of the magnetic field surrounding the wire. High current flow is caused by high voltage or low resistance. Conversely, low current flow results from high resistance or low voltage. Because current is the supply to get any electrical job done, an ammeter comes in handy for many types of diagnosis. Here are some examples of things you would use an ammeter to measure:

Component	Why Use the Ammeter
Generator/Alternator	The vehicle's charging system is the power station for everything electrical. Generators are matched to the total load the vehicle might generate so that it can maintain adequate voltage while supplying enough current. Use an ammeter, usually in conjunction with a voltmeter, to measure the ability of the generator to meet its intended output specification.
Battery	The battery stores current as potential under a certain pressure (voltage). Use an ammeter, usually in conjunction with a voltmeter, to measure the potential of the battery when it is working to see if it meets its intended load specification.
High-Power Circuits	Electric fuel pumps, HVAC blower motors, starters, electric cooling fans, window motors, solenoids, and any other high-load item can be tested in the same way to determine if they require more current than they were designed to handle. In contrast to the battery and the generator, these items consume the power generated by the charging system. In many cases, you can use an ammeter to not only find a component that has an electrical issue, but also to find out if the system it operates has a mechanical issue. An example would be a window motor that blows fuses due to an excessive demand for current. It could be that the motor is good but that the window is stuck and will not move freely causing high current demand on a circuit. A sharp technician must keep this in mind before replacing a component that is not faulty. Some internal engine problems are missed until after that starter that needed 600 amps to turn the engine over was replaced.

Task A.4   Check continuity and resistances in electrical/electronic circuits and components; interpret readings and determine needed repairs.

An ohmmeter has an internal power source. Meter damage may result if it is connected to a live circuit. Most currently produced ohmmeters are auto-ranging (automatically selecting the proper resistance range). If not, be sure to select the proper range when measuring a circuit. For example, when a component has resistance of 10,000 ohms, select the X1,000 or 10K range.

Some circuits contain components that have published resistance specs. This is where an ohmmeter comes in handy. Fuel injectors, solenoids, and sensors are examples of items that often have specifications for resistance. They can be a good way to determine if a component has a problem. If for example, a fuel injector has a specification of 16 ohms of resistance for its coil, and you find when measuring that it is significantly lower at 9 ohms, you can apply Ohm's law to determine that the injector is going to have a higher current demand from the component that drives it. In this case, it is driven by the PCM (powertrain control module). If it is demanding more current, you can also determine that since the injector driver in the PCM takes charging system current and supplies it to the injectors, the driver might start to get overloaded since it is a solid-state component looking to work in a tight set of conditions. The opposite is also true. If the resistance was too high, there might not be enough current passing through the injector to activate it. You can apply this to anything that has windings or wires. With that said, remember that resistance in an unloaded circuit is not the final word in diagnosis. You might find that a circuit with no load on it has very acceptable resistance; however, when there is a load on it, the resistance increases causing voltage drops and current losses.

Task A.5   Check electronic circuit waveforms; interpret readings and determine needed repairs.

An oscilloscope converts electrical signals into a visual image representing voltage changes over a specific period of time. An upward movement of the trace means the voltage has increased; whereas a downward movement means the voltage has decreased. If the trace stays flat, the voltage is staying at that level. As the trace moves across the screen of the oscilloscope, time is represented.

The size and clarity of the trace is dependent on the voltage scale and the time reference selected by the technician. Most scopes are equipped with controls that allow voltage and time interval selection.

Analog scopes show the actual activity of a circuit and are referred to as "real-time" scopes. Digital scopes convert the voltage signals into digital signals; therefore, some delay between the electrical activity and the display is experienced. This delay does give the trace a cleaner appearance, as an analog trace is only clean when the voltage has been constant for some time. Examine an example of a waveform captured from a digital oscilloscope.

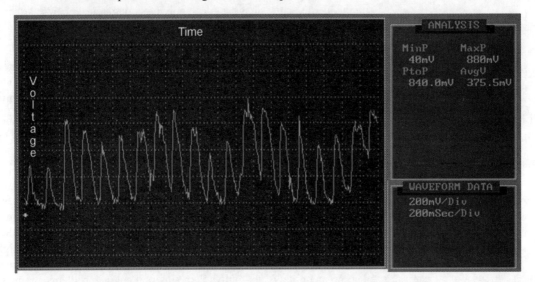

It is beyond the scope of this guide to teach you to use a labscope, so you must take this task on yourself. We can teach you to read a screen on a labscope so that you can answer questions on the ASE A6 test relating to them. This example is of an oxygen sensor, and it is a good oxygen sensor pattern. If you look at the screen, the horizontal plane is time. If we look at the area that says "waveform data," is 200 mV. A millivolt (mV) is one thousandth of a volt; so 1,000 mV equals 1volt. Each block across the screen represents a sample of time of 200mV. You can adjust the "time base" of the scope to better view faster or slower signals. As you can see by looking at the way the voltage of the signal goes up and down in the vertical plane, the oxygen sensor is switching about every 200 milliseconds or about every quarter second. A voltmeter would not be able to show you this detail, which might cause you to miss a problem that occurs very fast. The labscope is the power tool of the diagnostic technician. We want to draw your attention to one other area of the screen: the "analysis" section. The MinP stands for the lowest voltage the sensor achieved. The MaxP represents the highest voltage achieved. PtoP is a measurement of the total amplitude point-to-point that was achieved. Since the signal never went to 0 volts, you subtract the minimum number from the maximum number, and you get the point-to-point measurement. AvgV tells you the average voltage. This is a very handy item when looking at an oxygen sensor because it tells you about where the vehicle's fuel system is biased. This was probably a light cruise where the vehicle was able to run somewhat on the lean side, which is evident because the voltage is below the approximate 500mV centerline of a 1,000-mV oxygen sensor. Labscopes are used to diagnose items with rapid changing loads and functions. You would probably not use a labscope to measure a light bulb because it does not change very much. You would use it to watch the waveform of a fuel pump, for instance, to determine if there is abnormal wear on the brushes; they create a "signature" waveform that can help a technician diagnose a fuel pump that may not be causing a problem at that particular time.

Task A.6   Use scan tool data to diagnose electronic systems; interpret readings and determine needed repairs.

A scan tool is designed to communicate with the vehicle's computer. Connected to the computer through a diagnostic connector, a scan tool can access trouble codes, run tests to check system operations, and monitor the activity of a system. Many scan tools have the ability to set up special functions within the vehicle's computer systems that help a technician test circuits or to activate them so that you can watch the scan tool data for changes. This is known as bi-directional communication. The body controls of most late-model vehicles can be accessed and even exercised by the scan tool. We will talk more about this when it is appropriate to a given system.

The data displayed on a scan tool should be compared to the expected or known good data tables given in the service manual for the vehicle being tested. Depending on the year of the vehicle and the system with which it is equipped, the trouble codes displayed on the scan tool may indicate a general problem area or may identify a specific part of a circuit. Observing the activity of various inputs and outputs can lead a technician to problems that may be related or unrelated to the computer system.

Task A.7   Check electrical/electronic circuits with jumper wires; determine needed repairs.

A fused jumper wire may be used to bypass a part of a circuit to locate a defect in the circuit. Note: Jumper wires are used to bypass conductors, not loads. If wiring is suspect, a supply can be brought to a component with jumper wires to see if it operates correctly. Jumper wires can also be used in some diagnostic routines to remove a component from a circuit to test for high resistance.

Task A.8   Find shorts, grounds, opens, and resistance problems in electrical/electronic circuits; determine needed repairs.

Finding these types of problems is simply a matter of applying what we have learned about Ohm's law. If a circuit has a short to either power or ground, we can use an ohmmeter to chase across circuits to find the place where the short occurs. Shorts to ground are usually easier to trace because they usually involve contact with the body by a power circuit. Shorts made to one power circuit by another power circuit can be hard to find. The difficulty is the second power circuit crosses the first, and they are supplying power to each other. An example of this is when a dual-filament light bulb fails and crosses the brake and parking light circuits. This often results in the dash lights and park lamps coming on whenever the brake is applied. It seems very difficult to find these problems, but, if you apply what you know to studying a wiring diagram of the offending circuits, you can almost always find some good candidates for testing before you begin cutting open wire looms looking for melted wires. When it comes to open circuits, circuits that are not making connections anymore, patience and a wiring diagram are the ticket to tracing them down. Here again, think logically. It is far more likely that an open circuit will occur inside a connection or plug, a wire harness that is moved constantly or a failed component, than it is that a wire is broken inside a fixed harness. That is not to say that it does not happen, but the smart technician starts with the likely before chasing the less likely. Your tool for these procedures is your DMM.

Resistance tests can help you find shorts and grounds while voltage drops can help you isolate sources of open circuits or circuits with high resistance. ASE tests will approach this subject most often by showing you a diagram, posing a problem, and then asking you which wire or component in the circuit might be causing it.

Task A.9   Measure and diagnose the cause(s) of abnormal key-off battery drain (parasitic draw); determine needed repairs.

A parasitic draw occurs when a component stays on after it should shut off or when a component that should stay on begins to demand excessive current. Alarm systems, OBD II evaporative systems, PCMs, clock and radio memory, and convenience lighting are examples of systems that may stay active all the time or for quite some time after the vehicle is turned off. Light switches, relays, and generator voltage regulators are some very common causes of parasitic drain that can kill the battery over a few hours or a few days. Often the problems are intermittent and require tenacity to locate.

Many car manufacturers recommend measuring battery drain with a tester switch connected in series at the negative battery terminal. Follow the drain test procedure in the vehicle manufacturer's service manual. A multimeter with a milliampere scale is connected in parallel to the tester switch. When the tester switch is open, any current drain from the battery must flow through the multimeter. Some computers require several minutes after the ignition switch is turned off before they enter sleep mode with a reduced current drain. Therefore, after the ignition switch is turned off and the tester switch is opened, wait for the specified time before recording the milliampere reading. Some vehicle manufacturers specify a maximum battery drain.

Task A.10   Inspect, test, and replace fusible links, circuit breakers, fuses, and other current-limiting devices.

Current limiting devices can usually be checked with a voltmeter if they are connected in the circuit. The device will have voltage on both sides when in the closed position. If the device is open, there will be voltage at the input side but not at the output side. Some devices are one-time-use (e.g., fusible links and fuses). Others can be reset either manually or reset themselves automatically. Circuit breakers will sometimes get weak over time and trip at a current flow which is less than their rated flow. When an ohmmeter is connected to a circuit breaker, fuse, or fuse link, the meter should read zero ohms if the component is working properly. An open circuit breaker, fuse, or fuse link causes an infinite ohmmeter reading. The current flow from an ohmmeter will not cause an automotive circuit breaker to open.

Task A.11   Read and interpret electrical schematic diagrams and symbols.

All components have a special shape so they can be identified around the world on schematics. Some common components are shown in the figure.

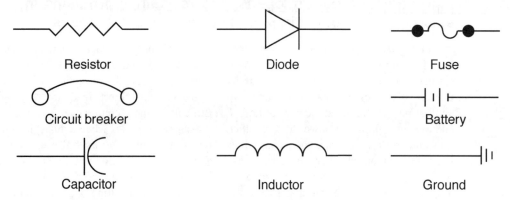

Resistor	Diode	Fuse
Circuit breaker		Battery
Capacitor	Inductor	Ground

B.   Battery Diagnosis and Service (4 Questions)

Task B.1   Perform battery state-of-charge test; determine needed service.

When performing a battery state of charge test with a hydrometer, you should subtract 0.004 specific gravity points from the hydrometer reading for every 10°F (5.6°C) of electrolyte temperature below 80°F (26.7°C). During this test, 0.004 specific gravity points must be added to the hydrometer reading for every 10°F (5.6°C) of electrolyte temperature above 80°F (26.7°C). The maximum variation is 0.050 in cell-specific gravity readings. When all the cell readings exceed 1.265, the battery is fully charged. Another common way to test the state of charge is with a voltmeter. After the battery has been stabilized (surface charge removed, if present), open circuit voltage should be 12.6 VDC.

Task B.2   Perform battery capacity (load, high rate discharge) test; determine needed service.

The battery discharge rate for a capacity test is usually one-half of the cold cranking rating. The battery is discharged at that rate for 15 seconds; the battery voltage must remain above 9.6V, with the battery temperature at 70°F (21.1°C) or above. The lower the temperature, the lower the voltage. After performing the initial test to remove the surface charge and allowing the battery to stabilize, it is often revealing to perform a second test. Many batteries fail the second test. If they fail, they are bad.

Task B.3   Maintain or restore electronic memory functions.

If battery voltage is disconnected from a computer, the adaptive memory in the computer is erased. In the case of a powertrain control module (PCM), disconnecting the power may cause erratic engine operation or erratic transmission shifting when the engine is restarted. After the vehicle is driven for about 20 miles (32 kilometers), the computer relearns the system, and normal operation is restored. If the vehicle is equipped with personalized items, such as memory seats or mirrors, the memory will be erased in the computer that controls these items. Radio station presets will also be erased. A 12-V power supply from a dry cell battery may be connected through the cigarette lighter or power point connector to maintain voltage to the electrical system when the battery is disconnected. It is very important that you check the information provided by the manufacturer before attaching a back-up power supply while the batter cables are removed. Some vehicles have been known to deploy the airbags. If in doubt, write down the vehicle radio presets and other presets to reprogram later and disconnect the battery negative post first.

Task B.4   Inspect, clean, fill, or replace battery.

A battery may be cleaned with any number of battery cleaning products. The primary purpose of cleaning a battery is to eliminate surface discharge across the top of the battery and to remove any corrosion that has collected around the terminals. Check and adjust the electrolyte level in batteries with removable caps.

Maintenance-free batteries with built-in hydrometers indicate a low electrolyte level when the hydrometer is light yellow or clear. If electrolyte is low, replace the battery. A low electrolyte level can be caused by a faulty voltage regulator that causes overcharging. When disconnecting battery cables always disconnect the negative cable first. Replacement of the battery is pretty straightforward, other than keeping in mind that cables and their connections must be clean to avoid voltage drops.

Task B.5   Perform slow/fast battery charge in accordance with manufacturer's recommendations.

If the battery is charged in the vehicle, disconnect the battery cables during the charging procedure. The charging time depends on the battery state of charge and the battery capacity. If the battery temperature exceeds 125°F (51.7°C) while charging, the battery may be damaged. When fast charging a battery, reduce the charging rate when specific gravity reaches 1.225 to avoid excessive battery gassing. The battery is fully charged when the specific gravity increases to 1.265. Do not attempt to fast charge a cold battery.

Task B.6   Inspect, clean, and repair or replace battery cables, connectors, clamps, and holddowns.

When removing battery clamps, take care not to stress the battery terminals. First, loosen the clamps and then remove them. If they do not come off readily, use a commercially available puller. Do not pry or apply sideways force to the terminals in order to avoid battery damage. Clean the terminal contact surfaces with a suitable tool until the surfaces are clean and bright, thus assuring good contact.

It is always wise to spray the cable clamps with a protective coating to prevent corrosion. Grease or petroleum jelly will also help prevent corrosion. Protective pads are available that go under the clamp and around the terminal to inhibit corrosion. If a cable has considerable corrosion, it might have corrosion that has worked up into the cable under the coating. When in doubt, do a voltage drop on the cables and recommend replacement if you find anything over 0.1 volt in the cable itself.

Task B.7   Jump start a vehicle with jumper cables and a booster battery or auxiliary power supply.

The accessories must be off in both vehicles during the boost procedure. The negative booster cable must be connected to an engine ground in the vehicle being boosted. Always connect the positive booster cable followed by the negative booster cable, and complete the negative cable connection last on the vehicle being boosted. Do not allow vehicles to contact each other. When disconnecting the booster cables, remove the negative booster cable first on the vehicle being boosted.

C. Starting System Diagnosis and Repair (5 Questions)

Task C.1   **Perform starter current draw test; determine needed repairs.**

A starter draw test is performed with a specialized ammeter that is generally also used for testing generators. Place an inductive clamp on the cable to the starter. Most inductive pickups are polarized which means that if you put the pickup on pointing in the wrong direction, you will get backward readings. Look for an arrow on the pickup. The arrow always points to the device you are testing. Disable the ignition system and either have an assistant crank the vehicle or use a remote starter button. There will be a slight spike in the current as the starter begins to crank. Unless this is significantly higher than the cranking amperage after the starter begins, it is not significant and is not the data you are seeking. In most applications, a starter will demand 130–200 amps, depending on the design of the starter and the size of the engine. Gear reduction starters do not take as much current to operate.

High starter current draw, low cranking speed, and low cranking voltage usually indicate a defective starter. This condition may also be caused by internal engine problems. Low current draw, low cranking speed, and high cranking voltage indicate excessive resistance in the starter circuit.

Task C.2   **Perform starter circuit voltage drop tests; determine needed repairs.**

Measure the voltage drop across each component in the starter circuit to check the resistance in that part of the circuit. The ignition and fuel system must be disabled while making these tests. Read the voltage drop across each component while the starting motor is operating. For example, connect the voltmeter leads to the positive battery terminal and the positive cable on the starter solenoid Crank the engine to measure the voltage drop across the positive battery cable. If you need help with this, refer back to our discussion of digital multimeters (DMM).

Task C.3   **Inspect, test, and repair or replace switches, connectors, and wires of starter control circuits.**

Relays and switches in the starting motor circuit may be tested with an ohmmeter. When an ohmmeter is connected across the relay or switch contacts, the meter should provide an infinite reading if the contacts are open. If the relay or switch contacts are closed, the ohmmeter reading should be at, or near, zero. When the ohmmeter leads are connected across the terminals that are connected to the relay winding, the meter should indicate the specified resistance. A resistance below the specified value indicates a shorted winding, whereas an infinite reading proves that the winding is open. Here again, if you suspect a fault in a switch, wire, or relay, perform a voltage drop test. In high amperage circuits, more than 0.1-volt drop may indicate a problem. Relays often develop corrosion on contacts which can cause problems with either power feed to the controlled device/system or activating the relay. Many relays have diodes in their activator circuits to avoid feedback. Electric motors can cause relays without diodes to stay engaged after the circuit is turned off. Starter solenoids are examples of large relays controlled by a fairly low amperage ignition switch. See Task E.1.4 for an explanation of relays.

Task C.4   **Inspect, test, and replace starter relays and solenoids.**

The ohmmeter leads must be connected across the solenoid terminal and the field coil terminal to test the pull-in winding. Connect the ohmmeter leads from the solenoid terminal to the ground to test the hold-in winding. See Task E.1.4 for an explanation of relays.

Task C.5   **Remove and replace starter.**

Replacing a starter motor is a very model-specific operation. You should always disconnect the negative battery cable before starting the procedure. There are few questions other than this that can be asked without favoring a specific brand. There may still be some questions out there about starter shims. The purpose of adding shims between the starter and its mounting surface is to adjust the depth, in one plane or the other, of the starter drive gear when it engages the flywheel. Virtually all vehicles have precision-engineered starter mounts that avoid the need for this. Remember, when replacing the starter, clean the mounting pad where the starter mounts up, as this is where many starters complete their ground circuit.

Task C.6   Differentiate between electrical and engine mechanical problems that cause a slow crank or no crank condition.

The first step to checking a starting problem should be to check the battery condition. Many starting problems can be attributed to battery problems. Do a complete series of battery tests to confirm the condition of the battery. Perform a visual inspection of the starting system to find loose connectors and frayed wires. After the visual inspection, if the engine will not turn over, try to turn the engine over manually by turning the crankshaft bolt. If the engine will not turn over but feels sort of elastic in response, at the crank bolt, remove the spark plugs and try again. You will probably find the engine has cylinders filled with coolant or fuel. If the engine feels very solid and will not turn over, try gently turning it backward. If it moves, you have a broken component stopping engine rotation. If you have no movement in either direction, there is probably a bearing failure seizing the engine. Slow cranking can be caused by internal failures. If the engine is cranking slow and starter current demand or draw is high, stop cranking and turn the engine manually to see if the problem is mechanical or simply a starter that has worn out.

D. Charging System Diagnosis and Repair (5 Questions)

Note: *In 1996 SAE J1930 terminology was adapted to standardized component identification. This standard adopted the name generator to refer to the component commonly known as an alternator. Both terms are used interchangeably in the ASE tests.*

Task D.1   Diagnose charging system problems that cause an undercharge, a no-charge, or an overcharge condition.

You can see in this chart that all three problems can be caused by different types of failures in the same components. Refer to the other tasks in this section for procedures to test the components and remember that many of these problems are circuit problems that can be traced by performing voltage drops on loaded circuits.

Condition	Possible Cause
Undercharging	• Loose generator belt
	• Faulty voltage regulator
	• Faulty generator
	• Poor connections at generator or battery
	• Faulty battery
No Charging	• Loose or missing generator belt
	• Faulty voltage regulator
	• Faulty generator
	• Poor connections at generator or battery
	• Faulty battery
Overcharging	• Faulty voltage regulator
	• Faulty generator
	• Poor connections at generator or battery
	• Faulty battery

Task D.2   Inspect, adjust, and replace generator (alternator) drive belts, pulleys, and tensioners.

An undercharged battery may be caused by a slipping generator belt. A slipping belt may be caused by insufficient belt tension or a worn, glazed, or oil-soaked belt. Belt tension may be tested with a belt tension gauge or by measuring the belt deflection in the center of the belt span. A belt should have 0.5 in (12.7 mm) of deflection for every foot (30.5 cm) of free span. Many serpentine belts have an automatic spring-loaded tensioner with a belt wear scale. These tensioners should be checked for movement by removing the belt and using the serpentine belt wrench to check for smooth full movement throughout the tensioner's complete operating range. If the tensioner has any sticking or binding spots, it should be replaced. When replacing the tensioner, be careful to clean the mounting surface properly. Dirt or other material at the mounting surface could cause the tensioner to be misaligned.

Task D.3   Perform charging system voltage output test; determine needed repairs.

Let's add a little more information to your pool of knowledge of generators. Automotive generators create alternating current (AC), which is converted to direct current (DC) output to be compliant with the rest of the vehicle's system. This function is performed by the rectifier. The power coming from the generator is 3-phase AC power. Each of the three wires reaching the rectifier bridge carries AC power, but in each wire the phase is different. It is the job of the rectifier to convert the power from 3-phase AC to single-phase DC. The DC power will not necessarily be at the correct voltage to charge the battery. The voltage regulator controls the field circuit in the generator to keep the output voltage as near to 14 volts as possible. One handy test that can be performed by most charging system testers or a DMM on the AC scale is to measure the AC voltage output at the output terminal of the generator. This should measure below 0.5 volts or less. The presence of AC voltage is an indication of problems with the rectifier or its diodes. Generator repair or replacement is required. Charging systems are tested for proper voltage output in the loaded and unloaded states.

Task D.4   Perform charging system current output test; determine needed repairs.

Most charging system tests are done with special charging system testers that have a device called a carbon pile that allows you to create a simulated load on the generator.

Before you begin any testing, you must determine what the rated output of the charging system is. In many vehicle lines, there is more than one available charging system. Often the only way to find the generators rated output is to find the stamping on the housing of the generator. These are more often than not impossible to see without a mirror or without removing the generator. Once that is determined, you can perform your tests by applying a load to the generator with all of the other vehicle accessories turned off. If the system cannot generate the specified output, you must determine if the cause is related to wiring issues or faults with the generator or voltage regulator. Each manufacturer has tests designed to reach this end, but a wiring diagram and DMM can be your most useful diagnostic tools. Normally a technician will check amperage output at about 2000 rpm.

Task D.5   Inspect, and test generator (alternator) control circuit; determine needed repairs.

The voltage regulator controls the field circuit in the generator to keep the output voltage as near to 14 volts as possible. In some applications, the voltage regulator is external of the generator and requires testing of wiring circuits before determining that there is a problem with the regulator. Many vehicles have internal or generator-mounted regulators. In most cases, if the regulator has failed, the generator will have sustained some damage as well. With modern generators creating as much as 150 amps, internal damage can occur quickly if the regulator fails. Before replacing a regulator, carefully inspect the generator and determine its age to avoid a comeback when the generator fails, often taking the new regulator with it. Still other modern generators are controlled by the BCM. Diagnosing these types of circuits will require the use of a scan tool.

Task D.6   Perform charging circuit voltage drop tests; determine needed repairs.

Connect a voltmeter from the generator battery wire to the positive battery terminal to measure voltage drop in the charging circuit. Many car manufacturers recommend a 10-amp charging rate while measuring this voltage drop. Other manufacturers recommend the charging system be charging at its maximum capacity. When the voltage drop is more than specified, the circuit resistance is excessive. High charging circuit resistance between the generator battery terminal and the positive battery terminal may cause an undercharged battery and an overworked generator.

Task D.7   **Inspect, repair, or replace connectors and wires of charging circuits.**

Always use Rosin core solder when repairing electrical circuits by soldering. Always insulate bare wires with heat shrink tubing. If the wire contains a shield wire, solder the drain wire separately. When using a heat and crimp seal, apply heat until the sealant appears at both ends. Butt (swage) connectors are suitable if care is taken to strip the proper amount of insulation from the wires. This will ensure adequate engagement in the connector without exposing the bare conductor. These connectors are not suitable in corrosion-prone areas or high amperage circuits.

Task D.8   **Remove, inspect, and replace generator (alternator).**

Disconnect the negative battery terminal before removing the generator. If the vehicle is equipped with an air bag system, wait for the time specified by the vehicle manufacturer, and then begin the generator removal procedure. Before you put the new generator in service, verify that the battery is fully charged.

E. Lighting Systems Diagnosis and Repair (6 Questions)

Task E.1   **Headlights, Parking Lights, Taillights, Dash Lights, and Courtesy Lights (3 Questions)**

Task E.1.1   **Diagnose the cause of brighter than normal, intermittent, dim, continuous, or no operation of headlights.**

Formerly one of the most basic circuits in a vehicle, the headlights have automatic dimming, light-sensing switches to automatically turn them on, and auxiliary driving lights integrated into them. Keep in mind that these items have their own controllers and relays that operate them. Always turn to your DMM and wiring diagrams to isolate problems. Ohm's law applies to every lighting circuit, too.

Concern	Possible Cause
Brighter than normal	• High charging system voltage
	• Dimmer switch in high beam position
	• Flash to pass or high beam relay stuck on
	• Bulb problems
	• Incorrect setting of automatic dimming sensor
Intermittent Operation	• Connection issues with switches, fire wall bulkhead connectors, in-line connectors, head lamp connectors, fuse, or relay blocks
	• Weak circuit breakers
	• Problems with head lamp relays
	• Shorts causing circuit breaker shutdown
	• Bulb problems
	• Faulty relays
	• Faulty switches (headlight or dimmer)

Concern	Possible Cause	*(Continued)*
Dim	• Poor grounds	
	• Contact corrosion in head lamp relays	
	• High resistance in connections at switches, fire wall bulkhead connectors, in-line connectors, head lamp connectors, fuse, or relay blocks	
	• Bulb problems	
No Headlights	• Bulb problems	
	• Open or high resistance connection issues with switches, fire wall bulkhead connectors, in-line connectors, head lamp connectors, fuse, or relay blocks	
	• Damaged circuit breakers or fusible links	
	• Faulty relays	
	• Faulty switches (headlight or dimmer)	
	• Shorts causing circuit breaker shutdown	

Task E.1.2   Inspect, test, and repair daytlme running light systems.

Daytime running lights (DRL) are typically part of the vehicle's high-beam circuit. The control circuit is connected directly to the vehicle's ignition switch so the lights are turned on whenever the vehicle is running. The circuit is equipped with a module that reduces battery voltage to approximately 6 volts. This voltage reduction allows the high beams to burn with less intensity and prolongs the life of the bulbs. When the headlight switch is moved to the ON position, the module is deactivated and the lights work normally.

Diagnosis of these systems should begin by identifying whether the problem is in the DRL system or the headlight system. If the problem is in the headlight system, service to the circuit and lamps is conducted in the same way as for vehicles not equipped with DRL. If the problem is in the DRL system and the headlights work normally, only that part of the circuit that is unique to the DRL can be the problem.

Task E.1.3   Inspect, replace, and aim headlights/bulbs and auxiliary lights (fog lights/driving lights).

When servicing the halogen or xenon bulbs used in head lamps and auxiliary lighting, always turn off the lights and allow the bulbs to cool, keep moisture away from the bulb, handle the bulb only by its base, do not scratch or drop the bulb, and coat the terminals of the bulb or the connector with dielectric grease to minimize corrosion.

Place sealed beam and standard bulbs securely into their retaining plates or fixtures. Aiming equipment includes a device that the lamp is aimed into while on, to set its height, boards with height marks on them, or devices that attach to the lamp to allow height and beam location to be set. Many lamps have built-in height adjustments.

Some composite lamps have numbers stamped on the lens that give the settings for alignment equipment that mounts on the headlamp alignment pads. When diagnosing auxiliary lighting systems, it is important to keep in mind that there are many different strategies for their function. Some will only work when the low beams are on and high beam settings may turn them off. The manufacturer's information and a wiring diagram are critical tools to successful solution of your customer's concern.

Task E.1.4   Inspect, test, and repair or replace headlight and dimmer switches, relays, control units, sensors, sockets, connectors, and wires of headlight circuits.

Some headlight switches have a button on the case that is designed to release the headlight knob. Some switches contain a circuit breaker that is connected to the headlamp system. Other lighting systems will have separate fuses. The dimmer switch in older applications is connected in series between the switch and the headlamps. When the current is turned on, the dimmer switch directs the current flow to the selected bulb. In the diagram you will see a very typical use of a relay.

Relays are employed by automotive engineers so that a low-amperage circuit can control a high-amperage device. Let us break down the function of this relay. Most relays have numbers on their terminals and most are standardized, thanks to the proliferation of Bosch relays throughout the automotive industry. The control circuit is made up of the 85 and 86 terminals. This is the low amperage side of the relay. In our example, you will notice that the ignition switch provides power through a fuse to terminal 86. This means our relay has power available whenever the ignition switch is turned on. It will not activate the relay until the switch on terminal 85 completes the ground. This causes a magnetic field to occur in the coil between 85 and 86, which pulls the contact to it and powers up the high amperage side of the circuit. If the relay had a diode, it would be between 85 and 86, clamping so that current could flow to ground but not back.

The next part of the relay we need to discuss is the high amperage side. In most applications, the 30 terminal is the power feed for the high amperage side of the circuit. In our example, it goes directly to the B+ side of the battery with a circuit breaker in-line. This circuit breaker might be in a fuse block or a separate component. Fuses and fusible links are often used to protect these circuits as well. You will see that there are two 87 terminals. Look closer, and you will see that one is 87 and the other is 87A. In this relay example, 87A is a tap to the unactivated position of the relay and the 87 terminal is the activated position. 87A is a handy terminal. Engineers use this terminal to reverse a circuit.

One example is the wide-open throttle cutout used on some air-conditioning compressors. The normal position is for the compressor's power to pass through the relay from 30 to 87A. When the PCM sees that there is a demand for power, it will activate the relay through pin 85 and 86, which will turn off the compressor clutch until demand comes back to normal. Since the majority of the time the compressor is not going to be turned off, the engineer takes the strategy LEAST likely to create component failure by using the inactivated position of the relay.

Now for the opposite side of that scenario: Consider a fuel pump relay application where the fuel pump is wired to terminal 87 (the activated side of the circuit). If 87A were used, the pump would run when the vehicle was turned off unless the relay were activated at shutdown. This would cause a parasitic drain eventually discharging the battery. When you are working with relays, review the circuit and follow the path of the power to determine what the circuit was designed to do.

When you are working with light control units and sensors, you MUST review the information the manufacturer provides. You may be tired of reading this, but it is absolutely the difference between successful electrical diagnosis and just plain hating every electrical problem that finds its way into your bay.

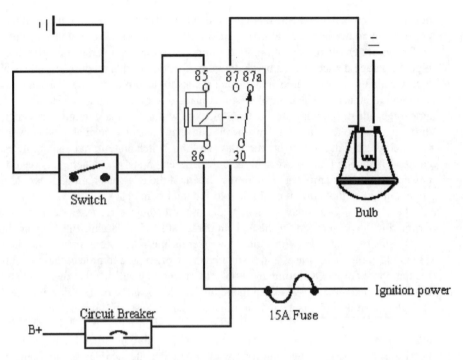

Task E.1.5   Diagnose the cause of brighter than normal, intermittent, dim, continuous, or no operation of parking lights, taillights, and/or auxiliary lights (fog lights/driving lights).

Some headlamp door retractors are vacuum operated. Any engine condition that would cause low vacuum will slow or prevent headlamp door operation. The vacuum system is generally used to close the headlamp door. Check for vacuum and wiring diagrams that show how the system works. Often the vacuum controls are electrically switched.

Electrically operated systems generally incorporate a headlamp control module that provides power to the headlamp door motors in response to signals received from the headlamp circuit. Probably the most common fault here is failure of the harnesses on the lamps from opening and closing.

Task E.1.6   Inspect, test, and repair or replace switches, relays, bulbs, sockets, connectors, wires, and controllers of parking lights, taillight circuits, and auxiliary light circuits (fog lights/driving lights).

When the driver turns the headlights on, voltage is supplied from the battery through the headlight switch to a terminal on the headlight door module. In response to this signal, the headlight door module supplies voltage to both headlight door motors. This action operates both motors to open the headlight doors. If the headlights are shut off, the module reverses the motor action to close the doors. Some headlight door motors have a manual knob on them. If the motors do not open the doors, this knob may be rotated to lift the doors. This is a typical scenario, but, due to the different ways manufacturers handle this type of system, you will MOST likely see very general questions.

Task E.1.7   Diagnose the cause of intermittent, dim, no lights, continuing operation, or no brightness control of instrument lighting circuits.

Be sure to take note of whether a problem is system-wide or only associated with one bulb. If a problem is limited to one lamp, your diagnostic approach is different than if it affects all of the lights in the circuit (or the body panel). In situations where only one is affected, you are looking for something that can only affect the one such as wires, bulbs, or a ground. In a circuit-wide problem, you are looking for the common areas, keeping in mind that sometimes every bulb in a circuit can be burned out by the time you see it for repair. If all lamps are out in an instrument panel, for instance, you might be looking for a common ground connection. When diagnosing an instrument cluster for intermittent problems, you should look for a common ground or power for the bulbs to see if it is loose.

Also inspect the printed circuit for cracks. If the instrument cluster is free of defects, check the circuit between the instrument cluster and the headlight switch for a damaged wire. Methods for diagnosis of other instrument panel lights are the same as instrument cluster lighting.

Concern	Possible Cause
Brighter than normal	• High charging system voltage
	• Bulb problems
Intermittent Operation	• Connection issues with switches, fire wall bulkhead connectors, in-line connectors, lamp connectors, fuse, or relay blocks
	• Weak circuit breakers
	• Problems with lamp relays
	• Shorts causing circuit breaker shutdown
	• Bulb problems
	• Faulty relays
	• Faulty switches (headlight or auxiliary)
	• Poor grounds
Dim	• Poor grounds
	• Contact corrosion in lamp relays/switches
	• High resistance in connections at switches, fire wall bulkhead connectors, in-line connectors, head lamp connectors fuse, or relay blocks
	• Bulb problems
No Lights	• Bulb problems
	• Open or high resistance connection issues with switches, fire wall bulkhead connectors, in-line connectors, fuse, or relay blocks
	• Damaged circuit breakers or fusible links
	• Faulty relays
	• Faulty switches (headlight or dimmer)
	• Shorts causing circuit breaker shutdown

Task E.1.8   Inspect, test, and repair or replace switches, relays, bulbs, sockets, connectors, wires, controllers, and printed circuit boards of instrument lighting circuits.

Bulbs in the instrument cluster are in parallel so that, if one bulb were to fail, the rest of the bulbs would still be able to illuminate. The rheostat in the headlight switch determines the brightness of the instrument cluster bulbs. If you go back to our discussion of Ohm's law, you will recall that we can vary voltage by adding or subtracting resistance in a circuit. This is the principle behind the rheostat. It is a variable resistor that dims the lights by putting more resistance in the instrument panel circuit.

A rheostat is connected in series with the instrument cluster bulbs. This rheostat is operated by the headlight switch knob or by a separate control knob. When the rheostat control knob is rotated, the voltage to the instrument cluster bulbs is reduced. This action lowers the current flow and reduces the brilliance of the bulbs. The instrument cluster bulbs are connected in parallel to the battery. If one bulb burns out, the other bulbs remain illuminated.

Task E.1.9   Diagnose the cause of intermittent, dim, continuous, or no operation of courtesy lights (dome, map, vanity, cargo, trunk, and hood light).

Courtesy lamps are normally controlled by the BCM on most late model vehicles. There are many different operating sequences for these lights. For example when the door is closed as the driver exists the vehicle many vehicles will leave the headlights, mirror, interior, and taillights on for a specified period of time. That time period may be reduced if the operator locks the doors upon existing. However if the trunk is opened upon existing the courtesy lights may be kept on for a longer period of time than normal, perhaps as long as twenty minutes. These lights are held on with the reasoning that the operator is still near the vehicle either loading or unloading the trunk. A soon as the trunk is closed , the normal timeout sequence is resumed by the BCM. If the hood is opened some vehicles will automatically turn on the parking lights. The reasoning is that the operator may be at the side of the road in an emergency situation and raising the hood. The parking lights are turned on to warn other motorists of the parked car. Be very careful when interpreting the owners concern with these types of lighting systems. What the owner feels is incorrect operation may be exactly what was designed into the system. Since these circuits are controlled by the BCM, diagnosing operation will likely involve using a scan tool.

Task E.1.10   Inspect, test, and repair or replace switches, relays, bulbs, sockets, connectors, wires, and controllers, of courtesy lights (dome, map, vanity, cargo, trunk, and hood lights) circuits.

Regardless of the light circuit being inspected, there are common things that need to be checked to identify the cause of operational problems. The entire light circuit should be visually inspected for loose and/or corroded connections. Likewise, the bulb terminals should also be inspected. These are all common sources of problems.

Naturally the cause of a light problem may be the lamp itself. It is safe to assume that when one bulb is not working, the bulb is bad. The problem can be verified by replacing the bulb with a new one. When doing this, make sure you install the same type bulb as the vehicle was originally equipped with, and make sure it is properly installed. If more than one bulb in a circuit is not working, it is very likely that the cause of the problem is not the bulbs. Check the system for an open. Light bulbs can be checked with an ohmmeter. If the meter reading across the terminals is infinite, an open (burned-out) bulb is indicated.

Task E.2   Stoplights, Turn Signals, Hazard Lights, and Backup Lights (3 Questions)

Task E.2.1   Diagnose the cause of intermittent, dim, continuous, or no operation of stoplights (brake lights).

Intermittent problems are typically caused by loose or poor connections. All connections must be clean and strong in order for all parts of a circuit to operate normally. High resistance in the circuit causes dim or poor operation of a lamp or other component. The most common cause for high resistance is corrosion. When a controller or switch fails to turn on or turn off a component, the switch is bad or there is a short in the control circuit or an open somewhere in the circuit. A quick voltage check at the component should help determine the cause. Once the cause is identified, the type of testing needed to further define the problem will be known.

In many stoplight circuits, voltage is supplied to the brake light switch from the fuse box. When the brakes are applied, brake pedal movement closes the stoplight switch. This action supplies voltage to the stoplights and the high stoplight. In many stoplight systems, the stoplight filaments are sharing the same light bulb encasement as the taillights.

Task E.2.2   Inspect, test, adjust, and repair or replace switches, bulbs, sockets, connectors, wires, and controllers of stoplight (brake light) circuits.

Regardless of the light circuit being inspected, there are common things that need to be checked to identify the cause of operational problems. Visually inspect the entire light circuit for loose and/or corroded connections. Likewise, inspect the bulb terminals. These are all common sources of problems.

Naturally, the cause of a light problem may be the lamp itself. It is safe to assume that when one bulb is not working, the bulb is bad. The problem can be verified by replacing the bulb with a new one. When doing this, make sure you properly install the same type bulb with which the vehicle was originally equipped. If more than one bulb in a circuit is not working, it is very likely that the cause of the problem is not the bulbs. Check the system for an open. Light bulbs can be checked with an ohmmeter. If the meter reading across the terminals is infinite, an open (burned-out) bulb is indicated.

The brake lights will not work on a vehicle. Everything else works normally.

Task E.2.3 Diagnose the cause of no turn signal and/or hazard lights, or lights with no flash on one or both sides.

The majority of late-model applications use electronic turn signal flashers. Unlike the older mechanical units that would slow down when a lamp burned out due to the decrease in current flow, electronic flashers and flasher relays either will not give the driver feedback on the instrument panel that a lamp is burned out, or possibly an open bulb could cause the flashing speed to increase.

In the case of the shared brake and turn lamp, when the brake pedal is applied during a right/left turn, the opposing brake light stays illuminated while the directional signal begins to flash. This is because the brake light circuit passes through the turn signal switch and can be overridden by the turn signals. When the hazard switch is pressed, voltage is applied to the hazard flasher through the hazard switch and signal light switch to the front and rear signal lights. Some vehicles have separate turn circuits that are amber in the rear. Their operation has no effect on the brake lamps. The hazard flasher has the same general design as conventional flashers. The hazard lamps are usually the turn signal lamps.

Task E.2.4 Inspect, test, and repair or replace switches, flasher units, bulbs, sockets, connectors, wires, and controllers of turn signal and hazard light circuits.

By now you can certainly see the pattern. If you understand how switches, lights, and Ohm's law works, all electrical circuits are pretty much the same.

Regardless of the light circuit being inspected, there are common things that need to be checked to identify the cause of operational problems. Visually inspect the entire light circuit for loose and/or corroded connections. Likewise, inspect the bulb terminals. These are all common sources of problems.

Naturally, the cause of a light problem may be the lamp itself. It is safe to assume that when one bulb is not working, the bulb is bad. The problem can be verified by replacing the bulb with a new one. When doing this, make sure you properly install the same type bulb with which the vehicle was originally equipped.. If more than one bulb in a circuit is not working, it is very likely that the cause of the problem is not the bulbs. Check the system for an open. Light bulbs can be checked with an ohmmeter. If the meter reading across the terminals is infinite, an open (burned-out) bulb is indicated.

Task E.2.5 Diagnose the cause of intermittent, dim, improper, continuous, or no operation of back-up lights.

With the addition of the neutral safety or back-up light switch, everything we have discussed in other areas of the lighting is true of back-up lamps as well. To review and expand, intermittent problems are typically caused by loose or poor connections. All connections must be clean and strong in order for all parts of a circuit to operate normally. Dim or poor operation of a lamp or other component is caused by high resistance in the circuit. The most common cause for high resistance is corrosion. When a controller or switch fails to turn on or turn off a component, the switch is bad or there is a short in the control circuit or an open somewhere in the circuit. A quick voltage check at the component should help determine the cause. Once the cause is identified, the type of testing needed to further define the problem will be known.

When the ignition switch is on and the back-up light switch is closed, voltage is supplied through these switches to the back-up lights. The gear selector linkage operates the back-up light switch. This switch might be mounted on the steering column, shifter, or transmission. The BCM will sometimes be used to control the backup lights. In this case diagnosis may very well involve testing with the scan tool.

Task E.2.6   Inspect, test, and repair or replace switch, bulbs, sockets, connectors, wires, and controllers of back-up light circuits.

Many back-up light switches are mounted on top of the steering column under the dash. These switches are often combined with the neutral safety switch. The combination back-up and neutral safety switch is operated by the gearshift tube in the steering column. The back-up light switch might be adjusted by loosening the mounting bolts and rotating the switch. Others are incorporated into the PRNDL switch mounted at the transmission.

Task E.2.7   Inspect, test, repair or replace trailer wiring harness and connector.

Many vehicles now are equipped with an optional trailer wiring harness. This harness allows the trailer lights to be controlled by the vehicle's light control circuit. In the past it was common for owners and aftermarket shops to install add-on trailer wiring connections. These connections simply added more lights in parallel to the lights on the vehicle. On older vehicles that was acceptable. Since many new vehicles now use LED lighting for stop, tail and turn lights, and add in most cases these lights use the BCM to operate. It is not acceptable to simply splice into the existing harness. And add this additional load.

F. Gauges, Warning Devices, and Driver Information Systems Diagnosis and Repair (6 Questions)

Note: Diagnosing causes of abnormal charging system gauge readings is limited to dash units and their electrical connections; other causes of abnormal charging system gauge readings are covered in category D.

Task F.1   Diagnose the cause of intermittent, high, low, or no gauge readings.

Many vehicles are equipped with thermal-electric gauges. These gauges contain a bimetallic strip surrounded by a heating coil. The pivoted gauge pointer is connected to the bimetallic strip. The sending unit contains a variable resistor. In a fuel gauge, this variable resistor is connected to a float in the fuel tank. Many late-model instrument panels incorporate electronic gauges that are driven by integrated circuits that take data lines from the PCM and convert them into analog gauge readings. In normal service, these units are serviced by specialists and not by technicians.

All gauge inputs come from some kind of sending unit. These are the most common failures and causes of inaccurate gauge readings. Because these units are almost all variable resistors, anything that alters resistance will cause incorrect gauge readings. Gauges that typically work this way are temperature, fuel, and oil pressure gauges.

Tachometers and electronic speedometers are examples of units that run off of a pulse, frequency, or voltage. Tachometers are driven by the negative side of the primary ignition and are calibrated for the number of cylinders in the engine. Connections are usually the only problems you see here. Most factory installed tachometers are ECM controlled. These tachometers are operated by a signal sent to the instrument panel controller. They can be tested for proper operation using a scan tool. Speedometers probably have the highest incidence of failure along with fuel and oil pressure gauges. Electronic speedometers are driven by a pulsed signal that is almost always an AC sine wave that comes from a signal generator attached to the transmission or differential. The speedometer is generally calibrated to expect a certain number of pulses per mile or is calibrated to the tire, transmission, and gear-ratio package of the vehicle. This information may be managed by the PCM and "translated" for the speedometer.

Methods of operating all gauges are as numerous as models of vehicles. Expect to demonstrate an understanding of how these gauges are driven on ASE tests, not the specifics of the design. Alternately, a design may be demonstrated to you along with some data that you need to determine the cause of a problem.

Task F.2   **Inspect, test, and repair or replace gauges, gauge sending units, connectors, wires, controllers, and printed circuit boards of gauge circuits.**

A typical gauge circuit is no more than a simple series circuit with a variable resistor. The variable resistor or sending unit responds to the change in fluid level or operating condition of the engine. Because the gauge is part of this series circuit, a change in circuit resistance will cause a change in voltage to the gauge and in current flow through the circuit. Unwanted resistance, from corroded terminals or similar problems, will cause the gauge to read incorrectly, as will high operating voltages and defective sending units. Sending units can be tested with an ohmmeter. Specifications are normally given for the sending unit in a variety of positions or conditions. Voltage checks at the gauge may also be necessary to diagnose the gauge circuit. Sending units, gauges, controllers, and printed circuit boards are replaced, not repaired, when they are found to be defective.

Task F.3   **Diagnose the cause(s) of intermittent, high, low, or no readings on electronic instrument clusters.**

Many electronic instrument displays provide an initial illumination of all segments when the ignition switch is turned on. This illumination proves the operation of the display segments. During this initial display, all the segments in the electronic instrument displays should be brightly illuminated for a few seconds. If some of the segments are not illuminated, replace the electronic instrument cluster. When none of the segments are illuminated, check the fuses and voltage supply to the display. Many electronic instrument displays have self-diagnostic capabilities. In some electronic instrument displays, a specific gauge illumination or digital display will indicate defects in the display. Other electronic displays may be diagnosed with a scan tool.

Task F.4   **Inspect, test, repair or replace sensors, sending units, connectors, wires, and controllers of electronic instrument circuits.**

Circuits for electronic instrumentation are very like those found in a conventional instrument circuit. When testing electronic instrument components, it is important to remember that electronic gauges can be either analog or digital. Analog gauges give the ability to show a constant change in value. Digital circuits operate in one of two states: on or off. The pulsing of the circuit (on and off) is what determines the readings on the instruments.

In order for an electronic gauge to display an accurate reading, it must receive an accurate signal from its sensor or sensors, it must receive the correct amount of voltage to operate, and it must function properly. Whenever diagnosing these systems, test the part of the circuit that is MOST likely to cause the observed problem.

Task F.5   **Diagnose the cause of constant, intermittent, or no operation of warning light, indicator lights, and other driver information systems.**

The MOST likely causes for constant operation of a warning or indicator light would be the existence of the situation the light was designed to warn the driver about or the control circuit is shorted and the switch or sender bypassed. If the warning light works intermittently, the MOST likely cause is a loose wire or connector causing an occasional short. If the light does not work at all, the bulb is burned out or there is an open in the circuit.

Task F.6   **Inspect, test, and repair or replace bulbs, sockets, connectors, wires, electronic components, and controllers of warning light, indicator light, and driver information system circuits.**

Some warning lights are operated by the body control module. The door ajar switches and the low washer fluid switch send an input signal to the body control module if a door is ajar or the washer fluid is low. When one of these signals is received, the body control unit grounds the appropriate circuit.

Some vehicles use a fiber-optic system to monitor lamp status. If a lamp is on, light will be transmitted through the fiber-optic cable to the display.

The driver information system is telling the operator that there is a stop light malfunctioning.

Task F.7   **Diagnose the cause of constant, intermittent, or no operation of audible warning devices.**

Most late-model vehicles that use a body control module use the inputs from door switches along with sensors in the ignition switch and seat belts to execute warning strategies. An example is if the key is in the ignition and the door is open with the engine off, a chime would sound. Another is if the vehicle is in motion or the transmission in gear and an occupied seat does not have a complete seat belt circuit, the warning lamp on the dash would illuminate and a chime would sound.

The body control units have many inputs including seat belts, door switches, wiper switch, horn, antitheft inputs, vehicle speed inputs, headlight switch, brake switch, and gauge warnings to touch on the most common. They use a series of outputs that include warning lights on the dash and different chimes/buzzers to let the driver know what is happening or to accomplish automatic functions like locking the doors from a vehicle speed or gear shift input. When diagnosing these, a scanner is almost always the quickest route. Beyond that, pinpoint tests can be performed with ohmmeters and voltmeters. Everything that applies to any other switch or output device is true here.

Task F.8   **Inspect, test, and repair or replace switches, relays, sensors, timers, electronic components, controllers, printed circuits, connectors, and wires of audible warning device circuits.**

Various types of tone generators, including buzzers, chimes, and voice synthesizers are used to remind drivers of a number of vehicle conditions. These tone generators should be checked for operation by running each through the prescribed self-test mode.

Audible warning devices are generally activated by the closing of a switch. A tone is emitted to warn the driver that something in the system is not functioning properly or that a situation exists that must be corrected. Many of the warning systems on today's vehicles are triggered by a PCM or body control unit and may be integrated into the unit. Always refer to the testing methods recommended by the manufacturer when testing these systems.

G. Horn and Wiper/Washer Diagnosis and Repair (3 Questions)

Task G.1   **Diagnose the cause of constant, intermittent, or no operation of horn(s).**

Many horn circuits utilize a relay. The voltage is supplied from the positive battery terminal through a fuse, circuit breaker, or fusible link to the relay winding and contacts. When the horn switch is closed on top of the steering column, the relay winding is grounded through the switch. This action closes the relay contacts, and voltage is supplied through these contacts to the horns. When the horn switch is closed in some circuits, voltage is supplied through the horn switch to the horns. A relay is not used in these circuits. Many vehicles have low and high pitch horns. Some horns have a pitch adjustment screw. Newer vehicles will have the horn relay operated by the BCM. This way the BCM can sound the horn as a vehicle theft deterrent and a signal to the operator for such things as door lock and unlock commands. Again, diagnosis of these systems will involve the use of a scan tool.

Task G.2   **Inspect, test, and repair or replace horn(s), horn relay, horn button (switch), connectors, wires, and controllers of horn circuits.**

Like any electrical circuit, the horn circuit and its components should be inspected for corrosion, damage, and looseness. Most vehicles produced since 1995 have air bags in the steering column. That means that an additional source of failure is present in the chain. The clock spring or spiral cable can be damaged. This will break the circuit from the horn button to the relay it activates. Many times, the vehicle will have an air bag light and/or cruise control problems at the same time since all functions of the steering wheel are transmitted by the clock spring. Take care when servicing horn contacts on air bag-equipped vehicles. Adhere to all safety standards. Weak horn operation is MOST likely caused by excessive resistance in the circuit.

Task G.3 Diagnose the cause of wiper problems including constant operation, intermittent operation, poor speed control, no parking, or no operation of wiper.

Some wiper motors contain a series field coil, a shunt field coil, and a relay. When the wiper switch is turned on, the relay winding is grounded through one set of switch contacts. This action closes the relay contacts, and current is supplied through these contacts to the series field coil and armature. Under this condition, the wiper motor starts turning. If the wiper switch is in the high-speed condition, the shunt coil is not grounded and the motor turns at high speed.

When the wiper switch is in the low-speed position, the shunt coil is grounded through the second set of wiper switch contacts. Under this condition, current flows through the shunt coil and the wiper switch to ground. Current flow through the shunt coil creates a strong magnetic field that induces more opposing voltage in the armature windings. This opposing voltage in the armature windings reduces current flow through the series coil and armature windings to slow the armature. If the wiper motor fails to park or parks in the wrong position, the parking switch or cam is probably defective.

Some wiper motors have permanent magnets in place of the field coils. These motors have a low-speed and a high-speed brush. In some of these motors, the low-speed brush is directly opposite the common brush, and the high-speed brush is positioned between these two brushes.

The most likely causes for constant operation of a wiper system would be a faulty or stuck switch or the controlling circuit is shorted and the switch is bypassed. If the wipers do not park properly, the park mechanism inside the wiper motor may be faulty, the park circuit damaged, or the intermittent wiper unit damaged. If the wipers work intermittently, the most likely cause is a loose wire or connector. If the wipers do not work at all, the wiper motor is bad or there is an open in the circuit.

Task G.4 Inspect, test, and replace intermittent (pulsing) wiper controls.

Many vehicles are equipped with intermittent or interval wiper motor circuits. In most of these circuits, the delay is set by the driver. A variable resistor in the intermittent wiper control module provides a voltage input to the intermittent wiper module. This module operates the wiper motor to provide the proper delay. In some intermittent wiper controls, various resistors are connected in the circuit as the switch is rotated.

When the wiper switch is placed in the high-speed position, voltage is supplied through the high-speed switch contact to the high-speed brush in the wiper motor. If the wiper switch is placed in the low-speed position, a voltage signal is sent from the wiper switch to the intermittent wiper module. When the signal is received, the module grounds the low-speed relay winding. This action closes the relay contacts, and voltage is supplied through these contacts to the low-speed brush in the wiper motor.

If the wiper switch is placed in one of the intermittent positions, a unique voltage signal is sent to the intermittent wiper module. When this signal is received, the module opens and closes the ground circuit on the low-speed relay winding to provide the proper delay interval.

Some vehicles are equipped with speed-sensitive wipers. This is a system that uses a speed signal from the vehicle speed sensor (VSS), either directly or from the PCM, to determine the speed of the wiper motor.

Task G.5 Inspect, test, and replace wiper motor, resistors, switches, relays, controllers, connections, and wires of wiper circuits.

The wiper circuit can be tested like any other circuit. Normally there are specifications available for available voltage to the wiper motor at various switch positions. The speed of the wiper motor is controlled by varying the voltage and current to the motor.

When checking the wiper system, never allow the wipers to run continuously across a dry windshield. The frictional drag on the glass may cause damage to the wiper motor. Wet the glass before operating the wipers.

Task G.6   Diagnose the cause of constant, intermittent, or no operation of window washer.

Many windshield washer systems have an electric pump mounted in the bottom of the washer fluid reservoir. When the washer button is pressed, voltage is supplied through the switch to the washer motor. This motor operates a pump that forces washer fluid through the hoses to the washer nozzles. Many late-model vehicles use the body control module to control the washers. They use a sequence when the washer is activated that activates the wipers and sprays the windshield for a certain number of cycles. In older cars that ran sequences, this was accomplished with cams and the washer unit physically attached to the wiper motor.

Task G.7   Inspect, test, and repair or replace washer motor, pump assembly, relays, switches, connectors, and wires of washer circuits.

When checking the windshield washer system, check for low fluid levels and disconnected wires. Then try to isolate the problem by disconnecting the hose at the pump and operating the system. If the pump ejects a stream of fluid, then the hoses are clogged. If the pump does not spray, observe the pump motor while activating the washer switch. If the motor operates, check for blockage at the pump. If there is no blockage, replace the motor. If the motor fails to operate, check for voltage and ground at the motor. This will isolate the problem to the motor or the washer switch and wires.

H.   Accessories Diagnosis and Repair (8 Questions)

Task H.1   Body (4 Questions)

Task H.1.1   Diagnose the cause of slow, intermittent, or no operation of power windows.

Power window circuits usually contain a master switch or an individual window switch and a window motor in each door. Some power window circuits have a window lockout switch to prevent passenger operation of the windows. When the master switch is placed in the down position, voltage is supplied to the center contact in this switch, through the movable switch contact to the brush on the lower side of the commutator.

The other brush is grounded through the master switch. Under this condition, the motor moves the window to the down position. When the up position is selected in the master switch, current flow through the motor is reversed. Voltage is supplied from the ignition switch circuit breaker and lockout switch to the window switch.

Task H.1.2   Inspect, test, and repair or replace regulators (linkages), switches, controllers, relays, motors, connectors, and wires of power window circuits.

Power window circuits usually include a circuit breaker at the motor. Because the motors are subject to high current flow, the breakers are present to prevent motor damage. A motor draws high current anytime it rotates slowly. The most common causes of a slow moving window are poor window track alignment, a buildup of ice on the window, or a weak motor.

The best way to diagnose a power window problem that is not caused by some physical interference or resistance is to refer to the wiring diagram for the circuit. Many switches are involved, and to quickly determine the problem you should identify all of the wires and connectors that are common to the system if all of the windows do not work properly. If only one window does not operate correctly, identify only those parts of the circuit that would affect that window.

Task H.1.3   Diagnose the cause of slow, intermittent, or no operation of power seat and driver memory controls.

A six-way power seat moves vertically at the front and rear and horizontally forward and rearward. This type of seat has two vertical motors and a horizontal motor. These motors are connected through gearboxes and cables to the seat track mechanisms. The front and rear switches have upward and downward positions, and the center switch has forward and rearward positions. When any of the switches are pressed, voltage is supplied to the appropriate motor in the proper direction. The motor moves the seat in the desired direction.

Task H.1.4 **Inspect, test, adjust, and repair or replace power seat gearbox, cables, switches, controllers, sensors, relays, solenoids, motors, connectors, and wires of power seat circuits and driver memory controls.**

Power seat circuits usually include a circuit breaker at the motors. Because the motors are subject to high current flow, the breakers are present to prevent motor damage. A motor draws high current anytime it rotates slowly. The most common causes of a slow moving seat are poor seat track alignment and a buildup of dirt on the tracks.

The best way to diagnose a power seat problem that is not caused by some physical interference or resistance is to refer to the wiring diagram for the circuit. Many switch positions and motors are involved, and to quickly determine the problem you should identify all of the wires and connectors that are common to the seat movement that is not working correctly. If none of the seat functions work, suspect an open to the control switch.

Task H.1.5 **Diagnose the cause of poor, intermittent, or no operation of rear window defogger.**

When the rear defogger switch is pressed, a signal is sent to a solid-state timer. When this signal is received, the timer grounds the relay winding. Under this condition, the relay supplies voltage to the defogger grid. When the relay is closed, current flows through the light emitting diode (LED) indicator to the ground. After a preset time, the timer opens the relay, shutting off the grid current.

Task H.1.6 **Inspect, test, and repair or replace switches, relays, timers, controllers, window grid, connectors, and wires of rear window defogger circuits.**

The grid tracks might be tested with a 12-V test light. As the test light is moved across a grid track from the power supply side to the ground side, the light should gradually become dimmer. If the test light goes out part way across the grid, the grid has an open circuit. A special compound is available to repair open circuits in the grid tracks.

Task H.1.7 **Diagnose the cause of poor, intermittent, or no operation of electric door and hatch/trunk lock.**

Most electric door lock circuits have small electric motors to operate the door locks. When either door lock switch is pushed to the lock position, voltage is supplied to all the door lock motors in the proper direction to provide lock action. If either door lock switch is pushed to the unlock position, voltage is supplied to all the door lock motors in the opposite direction to provide unlock action.

Task H.1.8 **Inspect, test, and repair or replace switches, relays, controllers, actuators/solenoids, connectors, and wires of electric door lock/hatch/trunk circuits.**

If all the door locks are completely inoperative, check the fuse. When all the door locks are inoperative in the lock or unlock mode, test the lock and unlock relays and connecting wires. If one door lock motor is inoperative, test the individual motor and connecting wire, and check the lock mechanism for a binding condition.

Task H.1.9 **Diagnose the cause of poor, intermittent, or no operation of keyless and remote lock/unlock devices.**

The remote keyless entry module is connected to the power door lock circuit. A small remote transmitter sends lock and unlock signals to this module when the appropriate buttons are pressed on the remote transmitter. When the hand-held remote transmitter is a short distance from the vehicle, the module responds to the transmitter signals. When the unlock button is pressed on the remote transmitter, the module supplies voltage to the unlock relay winding to close these relay contacts and move the door lock motors to the unlock position.

When the unlock button is depressed on the remote transmitter, the locks will unlock and the interior lights will illuminate on most systems. Then the remote keyless entry module will turn off the interior lights after approximately one minute or when the ignition is turned on.

Task H.1.10   Inspect, test, and repair or replace components, connectors, controllers, and wires of keyless and remote lock/unlock device circuits; reprogram system.

Power lock circuits usually control the action of a solenoid or motor. The movement of the center core of the solenoid controls the action of the lock levers and arms. The movement of the motor causes the lock mechanism to operate. The best way to diagnose a power lock problem is to refer to the wiring diagram for the circuit. Many switches and solenoids are involved, and to quickly determine the problem you should identify all of the wires and connectors that are common to the lock that is not working properly. If none of the locks work, suspect an open to the control switch. Some vehicles offer the customer the option to program the mode of operation of the door locks. For example, the customer can program one or all the door locks to unlock when the key is removed from the ignition. Additionally, the customer can program the door locks to unlock one or all the doors with one push of the button on the remote. Diagnosis of these systems will definitively involve understanding the customer's perception of how the system is supposed to work.

Task H.1.11   Diagnose the cause of slow, intermittent, or no operation of electrical sunroof and convertible/retractable top.

When the open sunroof switch is pressed, the open relay winding is grounded through the switch contacts. Under this condition, the relay contacts close and supply voltage to the sunroof motor brush. The other motor brush is connected through the close relay contacts to the ground. Current now flows through the motor, and the motor opens the sunroof. If the close button is pressed, the close relay winding is grounded through the close switch contacts. Under this condition, the close relay contacts supply voltage to the sunroof motor in the opposite direction to close the sunroof.

The convertible top system contains a dual switch, pump motor, hydraulic cylinders, and linkages from these cylinders to the convertible top. When the down button is pressed, voltage is supplied through these switch contacts to a motor brush. The opposite motor brush is grounded through the up contacts. Under this condition, current flows through the motor, and the motor drives the pump. With this motor rotation, the pump supplies hydraulic pressure to the proper side of the cylinder pistons to move the top downward. If the up button is pressed, motor and pump rotation are reversed and the pump supplies hydraulic pressure to the upward side of the cylinder pistons.

Task H.1.12   Inspect, test, and repair or replace motors, switches, controllers, relays, connectors, and wires of electrically-operated sunroof and convertible/retractable top circuits.

Power top and sunroof circuits are subject to high current flow. A motor draws high current anytime it rotates slowly. The most common causes of a slow moving top or sunroof motor are mechanical binding of the lift mechanisms or poor track alignments.

This type of mechanical interference can cause a motor to burn up, a circuit breaker to repeatedly trip, or damage to the circuit's wiring. The best way to diagnose a power top or sunroof problem that is not caused by some physical interference or resistance is to refer to the wiring diagram for the circuit. The control of the top or sunroof is based on the direction of current flow to the motors. Identify what parts of the circuit are involved with the function that is not working properly and diagnose those switches, wires, and connectors. If the sunroof or top does not move in any direction, suspect an open in the control circuit. The circuit breaker for a power convertible top opens when the top is raised.

Task H.1.13   Diagnose the cause of poor, intermittent, or no operation of electrically operated/heated mirror.

Voltage is supplied through a fuse to the power mirror switch assembly. When the mirror select switch is in the left position, it supplies voltage to the left mirror motor. When the left/right switch is pressed to the left position, a ground connection is completed from the left/right motor through the switch to the ground. Under this condition, the motor moves the left side mirror to the left. Similar action happens for all other directions and the right side mirror.

Task H.1.14   Inspect, test, and repair or replace motors, heated mirror grids, switches, controllers, relays, connectors, and wires of electrically operated/heated mirror circuits.

Some vehicles have heated mirrors. When the rear defogger button is pressed, the timer relay supplies voltage to the rear defogger grid and also to the heated mirror element. After 10 to 20 minutes, the timer relay shuts off the voltage supply to the defogger grid and the heated mirror element.

Task H.2   Miscellaneous (4 Questions)

Task H.2.1   Diagnose the cause of poor sound quality, noisy, erratic, intermittent, or no operation of the audio system; remove and reinstall audio system component (unit).

Audio systems typically work well, are noisy, or do not work at all. Since these units are typically replaced rather than repaired, a technician simply identifies the faulty part and replaces it. If the unit does not work at all, the problem is MOST likely a lack of power to the unit or a poor ground.

Sound quality depends on a number of things. Rattles and buzzes are caused more often by loose speakers, speaker mountings, speaker grilles, or trim panels than by inoperative speakers. Check the tightness of all mounting and trim pieces when this type of noise is heard.

Sound distortion can be caused by the speaker, radio chassis, or wiring. If the concern is the chassis, all speakers on the same side of the vehicle will exhibit the same poor quality. Distortion can also be caused by damaged wiring, which is normally accompanied by lower than normal sound output.

Static may be caused by the charging system or the ignition system. A poor engine ground or poor ground at the sound system components may cause static in the sound. Defective radio suppression devices, such as a suppression coil on an instrument voltage limiter or a clamping diode on an electromagnetic clutch, may cause static on the radio. A defective antenna with poor ground shielding may also result in static.

Task H.2.2   Inspect, test, and repair or replace speakers, amplifiers, remote controls, antennas, leads, grounds, connectors, and wires of sound system circuits.

An antenna may be tested with an ohmmeter. Continuity should be present between the end of the antenna mast and the center pin on the lead-in wire. Continuity also should be present between the ground shell on the lead-in wire and the antenna mounting hardware. No continuity should exist between the center pin on the lead-in wire and the ground shell. Speakers can be checked with an ohmmeter. On newer vehicles, the speakers can also be tested using a scan tool connected to the BCM.

Task H.2.3   Inspect, test, and repair or replace switches, relays, motor, connectors, and wires of power antenna circuits.

When the radio is turned on, voltage is supplied to the relay winding. This action moves the relay points to the up position, and current flows through the motor to move the antenna upward. When the antenna is fully extended, the up limit switch opens and stops the current flow through the motor.

When the radio is turned off, current flow through the relay coil stops. Under this condition, the relay contacts move to the down position. This action reverses current flow through the motor and moves the antenna downward. When the antenna is fully retracted, the down limit switch opens and stops the current flow through the motor.

An electric or electronic component with a varying magnetic field may cause radio static. A radio choke coil is connected to some components to reduce radio static. In some circuits, a radio suppression capacitor may be connected from the circuit to the ground to reduce radio static. An ohmmeter may be connected from the capacitor lead to the case to check the capacitor for insulation leakage between the capacitor plates. A capacitor tester may be used to test the capacitor for leakage, capacity, and resistance.

Task H.2.4   Inspect, test, and replace noise suppression components.

If the stator, a diode in the generator, or a noise suppression capacitor is defective, it could be the source of engine noise. An open field winding would cause the generator not to function. A defective generator will usually cause a whining noise. A defective spark plug will cause a snapping-type static.

Task H.2.5   Inspect, test, and repair or replace case, fuse, connectors, relays, and wires of cigar lighter/power outlet circuits.

Voltage is supplied from the positive battery terminal through a fuse to one terminal on the cigar lighter. The other terminal on the cigar lighter is connected to the ground. When the cigar lighter element is pushed inward, the circuit is completed through the lighter to the ground. Current flow through the lighter heats the lighter element. When the element is hot, the lighter element moves outward and opens the circuit. On some vehicles, the lighter fuse also supplies voltage to the dome light.

Task H.2.6   Inspect, test, and repair or replace clock, connectors, and wires of clock circuits.

Voltage is supplied from the positive battery terminal through a 10A fuse to the clock. The other clock terminal is grounded. The illumination control contains a variable resistor to control illumination brightness in the clock display. Today's digital clock has very little power consumption so there is no need to disconnect the clock for any reason.

Task H.2.7   Diagnose the cause of unregulated, intermittent, or no operation of cruise control.

Cruise control systems are often a combination of vacuum-operated devices, mechanical linkages, and electrical components. Always check for mechanical binding of the linkage before testing the vacuum and electrical components. Often there is a specified adjustment for the linkage. Check and adjust the linkage before doing more diagnostics. The vacuum devices can be checked for operation and leaks with a hand-held vacuum pump. When vacuum is applied, a diaphragm should move. The unit should be able to hold any applied vacuum for quite some time. Once the vacuum is released, the diaphragm should relax and return to its off position.

In some systems, the control module and the stepper motor are combined in one unit. A cable is connected from the stepper motor to the throttle linkage. The control unit receives inputs from the cruise control switch, brake switch, and vehicle speed sensor (VSS). The control module sends output commands to the stepper motor to provide the desired throttle opening. A defective VSS might cause erratic or no cruise control operation.

Many newer vehicles actuate an electric servo motor with a cable attached to the throttle body or linkage to control speed. The speed is controlled by the PCM. Inputs such as VSS, brake switch, steering wheel inputs, cruise control set and on/off switches all feed information to the PCM which responds by opening the throttle and monitoring the speed through the VSS. Keep in mind that most late-model systems use the clock spring or spiral cable to transfer the commands from steering wheel-mounted switches. Many times a vehicle with inoperative cruise control will also have the air bag light on as well if damage to the clock spring has occurred.

Task H.2.8   Inspect, test, adjust, and repair or replace regulator, servo, hoses, switches, relays, electronic controller, speed sensors, connectors, and wires of cruise control circuits.

Many vehicles have an electronic cruise control. In some of these systems, the control module and the stepper motor are combined in one unit. A cable is connected from the stepper motor to the throttle linkage. The control unit receives inputs from the cruise control switch, brake switch, and VSS. The control module sends output commands to the stepper motor to provide the desired throttle opening. A defective VSS might cause erratic or no cruise control operation. If a cruise control cable adjustment is required on these systems, remove the cruise control cable from the throttle linkage. With the throttle closed and the cable pulled all the way outward, install the cable on the throttle linkage. Turn the adjuster screw on the cruise control cable to obtain a 0.0197-in (5 mm) lash in the cable.

Some cruise control systems have the control module mounted in the PCM. The control module is connected to an external servo. This servo contains a vacuum diaphragm that is connected by a cable to the throttle linkage. The servo also contains a vent solenoid and a vacuum solenoid. The control module receives the same inputs as described previously. In response to these inputs, the control module operates the vent and vacuum solenoids to supply the proper vacuum to the servo diaphragm. Because the servo diaphragm is connected to the throttle, the vacuum supplied to this diaphragm provides the desired throttle opening. In these systems, a leak in the servo diaphragm might cause erratic cruise control operation or gradual reduction in the cruise set speed.

Task H.2.9   Diagnose the cause of false, intermittent, or no operation of antitheft system.

Most common false alarms are caused by misplaced sensors or overly adjusted sensors, such as shock sensors. Most new shock and glass sensors now have two-stage mechanisms where the sensor will give a warning when the first threshold is broken and will sound the alarm when the second threshold is broken. Door sensors will start to set false signals to the alarm module if they become rusted out or moving parts begin to wear out.

Many alarm systems tie into the interior dome light circuit to know when a door has been opened. Some alarm systems are tied directly into the door ajar switch or have a switch in the door for specifically detecting when the door is open. The reason for a separate switch is because some manufacturers have door handle switches, instead of door ajar switches that tell a control module to illuminate the interior lights.

Modern factory-installed antitheft systems use a dedicated control module or the functions are designed into an existing module. Most factory-installed systems are a passive-only system, which means the system is designed only to start the car when the correct key is inserted into the ignition. The system will disable the starter system, fuel delivery system, ignition system, or any or all of these systems if the right ignition key is not used. Active antitheft systems usually refer to a system that arms when the system is armed or when a sequence of events happens that automatically arm the system. An active system will sound the vehicle's horn(s) and disable the starter if it detects an attempt to break into one of its coverage zones. These zones can include the doors, trunk, hood, ignition, and a radio input. Some systems have active ultrasonic sound waves that set up an invisible shield to protect the interiors of convertibles.

Task H.2.10 Inspect, test, and repair or replace components, controllers, switches, relays, connectors, sensors, and wires of anti-theft system circuits.

An antitheft system is normally concealed within the vehicle and therefore it is difficult to visually inspect most of the components. Two important tools will allow you to properly test the system: a wiring diagram and a parts locator guide. Make sure you treat the circuit just like any other electrical circuit and look for the type of problem before proceeding. A short will cause the fuse of the circuit breaker to blow. An open will prevent operation of the circuit. High resistance will cause the system to work improperly. All systems are unique and have a variety of sensors and controllers. Make sure you understand how the system should work before proceeding.

Task H.2.11 Diagnose the cause(s) of the supplemental restraint/airbag warning light staying on or flashing.

In many air bag systems, the warning light is illuminated for five to six seconds after the engine is started while the module performs system checks. The air bag warning light will then turn off if the system passes all its system tests. If the warning light stays illuminated, then there is a fault with the air bag system. When the air bag module is set to diagnostic mode, a code will be flashed through the air bag warning light. If this light should also fail, then the system may be set up to sound a warning tone through the warning chime system. You should also be able to retrieve codes for the system with a diagnostic tool.

Task H.2.12 Disarm and enable the airbag system for vehicle service following manufacturer's recommended procedures.

The exact procedure for disarming an air bag system varies from vehicle to vehicle; always refer to the appropriate service manual to do this operation. Typically the procedure involves disconnecting the negative battery cable and taping the cable terminal to prevent accidental connection to the battery post. Then the SIR fuse is removed from the fuse box. After this, you need to wait at least ten minutes to allow the reserve energy to dissipate before working on or around the air bag system.

Task H.2.13 Inspect, test, repair or replace the airbag(s), controller, sensors, connectors, and wires of the airbag system circuit(s).

When servicing the air bag system, always disconnect the battery negative terminal first and wait for the manufacturer's specified time period to elapse. This time period is usually one or two minutes. Never use a powered test light to diagnose an air bag system. Diagnose these systems with a voltmeter or the manufacturer's recommended diagnostic tool(s). Use of an ohmmeter should be restricted to circuits without connections to pyrotechnic devices. Because deployed air bags may contain residual chemicals, wear safety glasses and gloves when handling these components.

Always mount sensors in their original orientation. Most sensors have directional arrows that must face the specified direction. Front air bag sensors are positioned toward the front of the vehicle and side air bag sensors are aimed toward the sides.

Always store inflator modules face up on the bench. Carry these components with the trim cover facing away from your body.

5

Sample Test for Practice

Sample Test

Please note the letter and number in parentheses following each question. They match the task in Section 4 that discusses the relevant subject matter. You may want to refer to the overview using the cross-referencing key to help with questions posing problems for you.

1. All of the following could cause a slow cranking condition **EXCEPT:**
   A. over-advanced ignition timing.
   B. shorted neutral safety switch.
   C. misaligned starter mounting.
   D. low battery state of charge. (C.6)

2. During an output test using the full-field method, a 100-ampere alternator with an integral regulator produces 60 amperes. The cause of the low alternator output could be a(n):
   A. shorted diode in the alternator.
   B. broken brush lead wire in the alternator.
   C. open circuit in the voltage regulator.
   D. defective alternator capacitor. (D.4)

3. A six-way power seat moves vertically at the front and rear, but there is no horizontal seat movement. All of these defects may be the cause of the problem **EXCEPT:**
   A. a newspaper jammed in the seat track mechanism.
   B. an open circuit between the switch and the horizontal motor.
   C. an open in the circuit from the switch assembly to ground.
   D. burned contacts in the horizontal seat switch. (H.1.4)

4. What type of problem is indicated by a lower than normal amperage reading when a circuit is activated?
   A. An open
   B. A short
   C. High resistance in the circuit
   D. A blown fuse (A.3)

5. The technician is having problems with the scan tool communicating with the BCM. Technician A says the ignition key must be on. Technician B says the engine must be running. Who is right?
   A. A only
   B. B only
   C. Both A and B
   D. Neither A nor B (A.6)

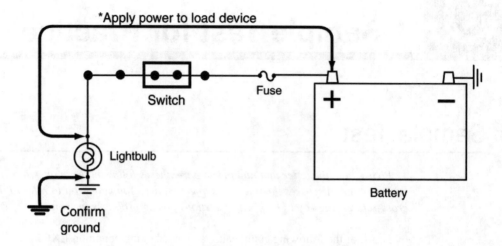

6. The light bulb is inoperative in the accompanying figure. A jumper wire is connected from the battery positive terminal to the light bulb with the switch on, and the bulb is not illuminated. With the switch in the on position, the bulb is illuminated when a jumper wire is connected from the ground side of the light bulb to ground. The cause of the inoperative light bulb could be a(n):
   A.   open circuit in the switch.
   B.   open circuit in the light bulb ground.
   C.   open circuit breaker.
   D.   burned-out fuse.   (A.7)

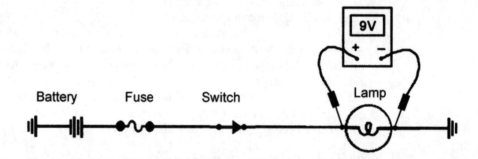

7. The battery in the drawing is fully charged and the switch is closed. The voltage drop across the light indicated on the voltmeter is 9V. Technician A says there may be a high resistance problem in the light. Technician B says the circuit may be grounded between the switch and the light. Who is right?
   A.   A only
   B.   B only
   C.   Both A and B
   D.   Neither A nor B   (A.2)

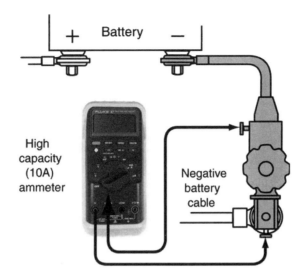

8. While performing the battery drain test in the figure,:
    A. the tester switch should be closed while starting or running the engine.
    B. a battery drain of 125 milliamperes is considered normal and will not discharge the battery.
    C. the actual battery drain is recorded immediately when the switch is opened.
    D. the driver door should be open while measuring the battery drain.   (A.9)

9. A circuit breaker is removed from a functioning power seat circuit, and an ohmmeter is connected to the circuit breaker terminals. Technician A says the ohmmeter should provide an infinite reading if the circuit breaker is satisfactory. Technician B says the ohmmeter current may cause the circuit breaker to open. Who is right?
    A. A only
    B. B only
    C. Both A and B
    D. Neither A nor B   (A.10)

10. A test light connected to ground does not illuminate when touched to either side of a fuse in the fuse box. Technician A says that the fuse is open. Technician B says that there is no power to the fuse. Who is right?
    A. A only
    B. B only
    C. Both A and B
    D. Neither A nor B   (A.10)

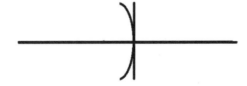

11. The symbol in the figure is a:
    A. resistor.
    B. capacitor.
    C. diode.
    D. inductor.   (A.11)

12. Technician A says that battery state of charge can be tested using a voltmeter. Technician B says that a hydrometer reading must be corrected based upon temperature. Who is right?
    A. A only
    B. B only
    C. Both A and B
    D. Neither A nor B   (B.1)

13. While discussing a battery capacity test with the battery temperature at 70°F (21.1°C), Technician A says the battery load rate is calculated by multiplying the battery reserve capacity rating times two. Technician B says the battery is satisfactory if the voltage remains above 9.6V throughout the test. Who is right?
    A. A only
    B. B only
    C. Both A and B
    D. Neither A nor B   (B.2)

14. The battery voltage is disconnected from the electrical system in a vehicle with many on-board computers. This procedure may cause:
    A. damage to all the computers.
    B. the engine to not start.
    C. erasure of the computers' adaptive memories.
    D. voltage surges in the electrical system.   (B.3)

15. Technician A says a shorted winding in an electrical component causes higher than normal current flow. Technician B says high resistance in a winding causes higher voltage drop across the winding. Who is right?
    A. A only
    B. B only
    C. Both A and B
    D. Neither A nor B   (A.8)

16. While testing a battery, a buildup of dirt and corrosion on the battery case, hold downs, and terminals is quite evident. Technician A says this could cause battery drain. Technician B says the battery must be leaking and should be replaced. Who is right?
    A. A only
    B. B only
    C. Both A and B
    D. Neither A nor B   (B.4)

17. While charging a battery at a high rate of current to fast charge the battery, Technician A connects a voltmeter across the battery to make sure the voltage does not exceed 15.5 volts. Technician B monitors the battery's temperature and is prepared to reduce the charging rate when the temperature approaches 125°F. Who is right?
    A. A only
    B. B only
    C. Both A and B
    D. Neither A nor B   (B.5)

18. All of the statements about servicing the battery and electrical system in an air bag-equipped vehicle are true EXCEPT:
    A. disconnect the negative battery cable first
    B. reconnect the positive battery cable first.
    C. begin servicing the electrical system immediately after the battery cables are disconnected.
    D. connect a 12V source to the cigarette lighter socket before disconnecting the battery cables.   (B.6)

19. While jump-starting a vehicle with a booster vehicle, Technician A says the accessories should be "on" in the booster vehicle while starting the vehicle being boosted. Technician B says the negative booster cable should be connected to an engine ground on the vehicle being boosted. Who is right?
    A. A only
    B. B only
    C. Both A and B
    D. Neither A nor B   (B.7)

20. An ohmmeter is used to test a fusible link. Technician A says voltage should be supplied to the fusible link during the ohmmeter test. Technician B says if the ohmmeter reading is 0 Ohms, the fusible link must be replaced. Who is right?
    A. A only
    B. B only
    C. Both A and B
    D. Neither A nor B   (A.10)

21. During a starter current draw test, the current draw is more than specified, and the cranking speed and battery voltages are less than specified. The cause of this problem may be:
    A. worn bushings in the starter motor.
    B. high resistance in the field windings.
    C. high resistance in the battery positive cable.
    D. a burned solenoid disc and terminals.   (C.1)

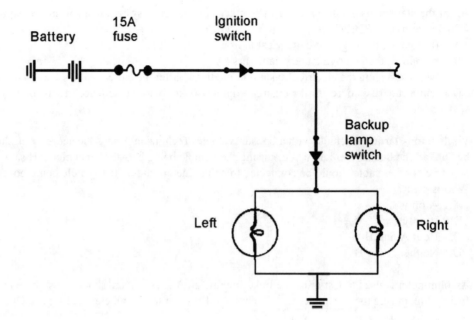

22. The right-hand back-up light circuit is grounded on the switch side of the bulb in the figure shown. Technician A says this condition may cause the back-up light fuse to fail. Technician B says the left-hand back-up light may work normally while the right-hand back-up light is inoperative. Who is right?
    A. A only
    B. B only
    C. Both A and B
    D. Neither A nor B   (E.2.6)

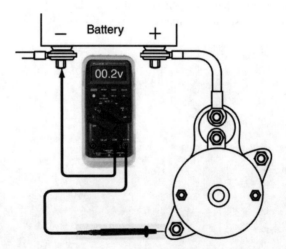

23. In the figure, the voltmeter is connected to test the voltage drop across the:
    A. positive battery cable.
    B. starter solenoid windings.
    C. starter ground circuit.
    D. starter solenoid disc and terminals.   (C.2)

24. The headlights on a vehicle go out intermittently and come back on in a few minutes. Technician A says this problem may be caused by an intermittent short to ground. Technician B says this problem may be caused by high charging system voltage. Who is correct?
    A. A only
    B. B only
    C. Both A and B
    D. Neither A nor B　　　　　　　　　　　　　　　　　　　　　　　　　(E.2.1)

25. When performing a voltage drop test on the negative (ground) side of the starter circuit, a reading of 1.5 volts is obtained. All of the following could cause this result **EXCEPT**:
    A. a corroded negative battery terminal.
    B. a loose ground connection at the motor.
    C. high resistance in the starter field windings.
    D. an improperly mounted starter.　　　　　　　　　　　　　　　　　　(C.2)

26. With the ignition switch off, the ohmmeter leads are connected from the starter solenoid S terminal to the solenoid case. The ohmmeter provides an infinite (∞) reading. The cause of this reading could be:
    A. the solenoid pull-in winding is grounded.
    B. the solenoid hold-in winding is open.
    C. the solenoid pull-in winding is shorted.
    D. the solenoid hold-in winding is shorted.　　　　　　　　　　　　　　(C.4)

27. A vehicle with a no crank condition is being tested. When the battery and start terminals on the starter's solenoid are connected with a jumper wire, the starter begins to crank the engine. Technician A says the starter solenoid may be bad. Technician B says the ignition switch may be bad. Who is right?
    A. A only
    B. B only
    C. Both A and B
    D. Neither A nor B　　　　　　　　　　　　　　　　　　　　　　　　　(C.4)

28. A customer complains that the engine will not turn over. Technician A says to first check if the engine will turn over by turning the crankshaft pulley nut. Technician B says to check the battery condition first. Who is right?
    A. A only
    B. B only
    C. Both A and B
    D. Neither A nor B　　　　　　　　　　　　　　　　　　　　　　　　　(C.6)

29. Technician A says if the charging system voltage is too high, there may be a loose or glazed drive belt. Technician B says if the charging system voltage is too low, the fault might be a grounded field wire from the regulator (full fielding the alternator). Who is right?
    A. A only
    B. B only
    C. Both A and B
    D. Neither A nor B　　　　　　　　　　　　　　　　　　　　　　(D.2 and D.3)

30. Charging system voltage is being measured at two places at the same time. With the voltmeter's leads across the battery, a 12.8-volt reading is observed. Connecting the voltmeter to the generator output terminal and the case results in a 14.2-volt reading. Technician A says there may be excessive resistance on the ground side of the charging system. Technician B says there may be excessive resistance on the positive side of the charging system. Who is right?
    A. A only
    B. B only
    C. Both A and B
    D. Neither A nor B   (D.6)

31. A fusible link at the alternator is suspected to be open. Technician A says that battery voltage should be present at both sides of the fusible link. Technician B says that a replacement fusible link should never be longer than 9 inches. Who is right?
    A. A only
    B. B only
    C. Both A and B
    D. Neither A nor B   (D.7)

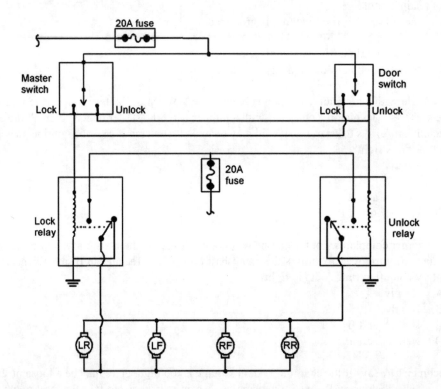

32. The RR door power lock is inoperative, but all the other power door locks operate normally as shown in the figure. Technician A says the door lock relay winding might have an open circuit. Technician B says the door unlock relay contacts might have an open circuit. Who is right?
    A. A only
    B. B only
    C. Both A and B
    D. Neither A nor B   (H.1.7)

33. The headlights on a vehicle go out intermittently and come back on in a few minutes. Technician A says this problem may be caused by an intermittent short to ground. Technician B says this problem may be caused by high charging system voltage. Who is right?
    A. A only
    B. B only
    C. Both A and B
    D. Neither A nor B   (E.1.1)

34. Which of the following is the LEAST likely cause for an inoperative retractable headlight door?
    A. A defective headlight motor
    B. Loose or broken vacuum lines
    C. A poor headlamp ground
    D. A faulty limit switch   (E.1.2)

35. All of these statements about halogen headlight bulb replacement are true **EXCEPT:**
    A. handle the halogen bulb only by the base.
    B. do not drop or scratch the bulb.
    C. change the bulb with the headlights on.
    D. keep moisture away from the bulb.   (E.1.3)

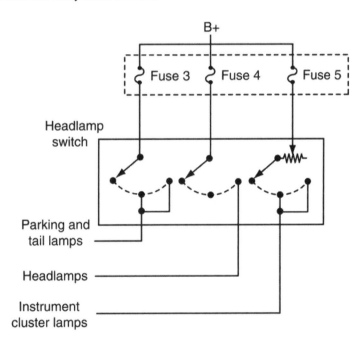

36. If fuse 5 in the circuit shown is open, what would happen?
    A. Inoperative taillights
    B. Inoperative stoplights
    C. Inoperative instrument cluster lights
    D. Inoperative low-beam headlights   (E.1.4)

37. An air bag warning light is illuminated intermittently with the engine running. Technician A says the air bag system has an electrical defect. Technician B says this defect may cause the air bag to inflate accidentally. Who is right?
    A. A only
    B. B only
    C. Both A and B
    D. Neither A nor B   (H.2.11)

38. In a headlight system equipped with doors that use separate electric motors and a control module, Technician A says some of these systems have a manual override knob to open the doors if the system fails. Technician B says the control module reverses polarity through the door motor to provide door opening and closing action. Who is right?
    A. A only
    B. B only
    C. Both A and B
    D. Neither A nor B   (E.1.5)

39. While diagnosing nonfunctional daytime running lights, Technician A, after some preliminary inspections, checks the circuit between the headlamp switch and the control module. Technician B begins the diagnosis with a check of the vehicle's high-beam lamps. Who is right?
    A. A only
    B. B only
    C. Both A and B
    D. Neither A nor B   (E.1.7)

40. While describing the typical procedures for troubleshooting a tone generator system on a late-model vehicle, Technician A says that if the device does not work at all, the sounding device is bad or there is an open in the circuit. Technician B says the tone generator should be checked for operation by running it through the prescribed self-test mode. Who is right?
    A. A only
    B. B only
    C. Both A and B
    D. Neither A nor B   (F.8)

41. A vehicle has one dim taillight and the other taillights have normal brilliance. Technician A says there might be high resistance between the dim taillight and the ground. Technician B says there might be a defect in the taillight contacts in the headlight switch. Who is right?
    A. A only
    B. B only
    C. Both A and B
    D. Neither A nor B   (E.1.8)

42. The washer nozzles do not spray fluid when the switch is activated. The pump does not operate. Technician A says the pump could be seized. Technician B says the nozzles could be plugged. Who is right?
    A. A only
    B. B only
    C. Both A and B
    D. Neither A nor B   (G.6)

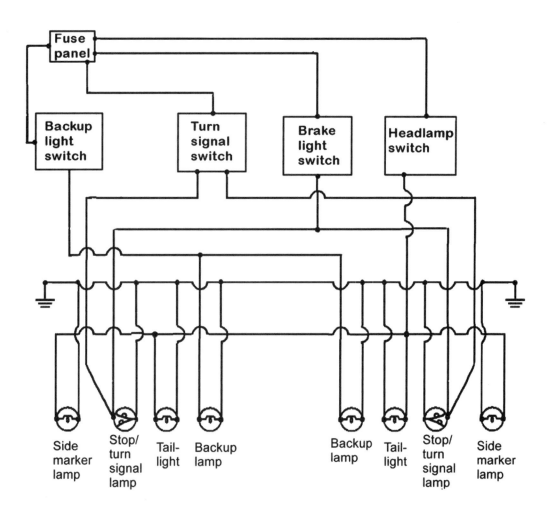

43. The rear light ground connection on the left side of the figure has an open circuit. The ground connection on the right side of the figure is satisfactory. This problem could result in:
    A. inoperative LR taillight, stoplight, and side marker lights.
    B. no change in operation of the rear lights.
    C. inoperative LR taillight and stoplights only.
    D. inoperative backup lights.   (E.1.8)

44. A windshield wiper system operates at low speed only regardless of switch position. Technician A says the problem may be the switch. Technician B says the cause of the problem may be a worn high-speed brush. Who is right?
    A. A only
    B. B only
    C. Both A and B
    D. Neither A nor B   (G.3)

45. While diagnosing an instrument panel lighting circuit, Technician A says the power source for the lamps is provided through the headlight switch and/or a rheostat. Technician B says printed circuit boards should be replaced if they are faulty and cannot be repaired. Who is right?
    A. A only
    B. B only
    C. Both A and B
    D. Neither A nor B   (E.10)

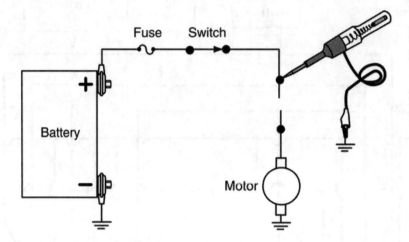

46. When an open circuit occurs at the connector in the drawing and 12V are supplied from the battery to the circuit:
    A. the test light is illuminated when connected as in the figure.
    B. as connected to the circuit in the figure, the test light will not illuminate.
    C. if the test light were connected to the ground of the motor, it would illuminate.
    D. if the test light were connected to the other side of the connector, it would illuminate. (A.1)

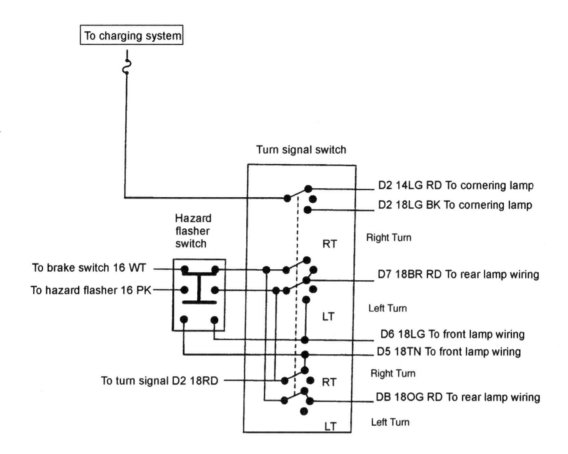

47. In the signal light circuit in the figure, the right rear signal light is dim and all the other lights work normally. The cause of this problem may be:
    A. high resistance in the DB 180G RD wire from the signal light switch to the rear lamp wiring.
    B. a short to ground in the DB 180G RD wire from the signal light switch to the rear lamp wiring.
    C. high resistance in the D7 18BR RD wire from the signal light switch to the rear lamp wiring.
    D. high resistance in the D2 18 RD wire from the signal light flasher to the switch.   (E.2.3)

48. The turn signals are not operating properly. Technician A says the multi-function switch can be checked with a pin-out chart and a DMM. Technician B says the problem could be a burned out bulb. Who is correct?
    A. A only
    B. B only
    C. Both A and B
    D. Neither A nor B   (E.2.4)

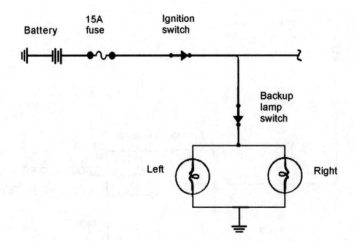

49. The back-up lights shown in the figure are inoperative. With the gear selector in reverse and the ignition switch on, the voltage measured on both sides of the back-up light bulbs is 12V. The cause of this problem could be:
    A.   shorted filaments in the back-up light bulbs.
    B.   an open circuit between the back-up lights and ground.
    C.   high resistance in the back-up light switch contacts.
    D.   the wire from the back-up switch to the bulbs is shorted to 12V.   (E.2.5)

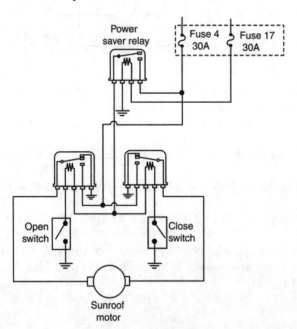

50. The power sunroof in the figure is completely inoperative. The cause of this problem could be an open circuit:
    A.   in the power-saver relay winding:
    B.   in the close relay winding.
    C.   at the right-hand contact in the close switch.
    D.   between the close relay winding and close switch.   (H.1.10)

51. Which of the following is the best method to test a trailer wiring connector?
   A. DMM
   B. Testlight
   C. Ohmmeter
   D. Analog voltmeter   (E.2.7)

52. A remote keyless entry system does not operate unless the person holding the remote control
   is right beside the vehicle. Technician A says to replace the body control module (BCM).
   Technician B says to replace the remote control door lock receiver. Who is correct?
   A. A only
   B. B only
   C. Both A and B
   D. Neither A nor B   (H.1.9)

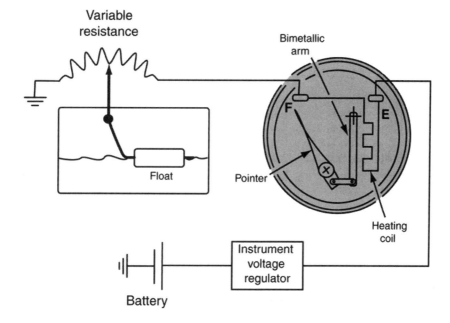

53. The fuel gauge in the figure reads full continuously when the ignition switch is on. All the other
   gauges operate normally. The cause of the problem might be:
   A. high resistance in the sending unit ground wire.
   B. high resistance between the instrument voltage regulator and the gauge.
   C. a short to ground between the gauge and the sending unit.
   D. an open circuit in the wire from the gauge to the sending unit.   (F.1)

54. All the gauges are erratic in an instrument panel with thermal-electric gauges and an instrument
   voltage limiter. Technician A says the alternator may be the problem. Technician B says the
   instrument voltage limiter may be the problem. Who is correct?
   A. A only
   B. B only
   C. Both A and B
   D. Neither A nor B   (F.2)

55. A power window operates normally from the master switch, but the window does not work using the window switch. The cause of this problem may be a(n):
    A. open circuit between the ignition switch and the window switch.
    B. open circuit in the window switch movable contacts.
    C. open circuit in the master switch ground wire.
    D. short to ground at the circuit breaker in the motor.   (H.1.1)

56. An electronic instrument cluster is completely inoperative. Technician A says to replace the instrument cluster. Technician B says to test the instrument panel fuse. Who is correct?
    A. A only
    B. B only
    C. Both A and B
    D. Neither A nor B   (F.3)

57. The left rear parking lamp fails to operate. The right rear parking lamp works normally. Technician A says the problem could be a worn socket on the right rear parking lamp. Technician B says the problem could be a faulty headlight switch. Who is correct?
    A. A only
    B. B only
    C. Both A and B
    D. Neither A nor B   (E.1.6)

58. The clock will not function on a customer's vehicle. Technician A says the problem could be in the BCM. Technician B says the problem could be with the clock fuse. Who is right?
    A. A only
    B. B only
    C. Both A and B
    D. Neither A nor B   (H.2.6)

59. A digital speedometer constantly reads zero mph. Technician A says a faulty speed sensor may cause the problem. Technician B says the cause may be an open throttle position sensor. Who is right?
    A. A only
    B. B only
    C. Both A and B
    D. Neither A nor B   (F.4)

60. All of the following can cause the vehicle's warning lights not to work **EXCEPT** a(n):
    A. burned-out bulb.
    B. short to ground in the sending unit circuit.
    C. open in the sending unit circuit.
    D. defective sending unit.   (F.5)

61. The driver information system is telling the operator that there is a stop light malfunctioning. Technician A says the problem could be a burned out bulb. Technician B says the problem could be the center high mounted stop light. Who is correct?
    A. A only
    B. B only
    C. Both A and B
    D. Neither A nor B   (F.6)

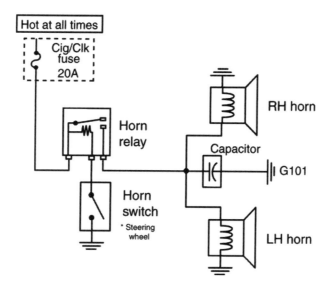

62. The horn in the figure sounds continuously. Technician A says the wire from the relay winding to the horn switch might be grounded. Technician B says the relay points might be stuck in the open position. Who is right?
    A. A only
    B. B only
    C. Both A and B
    D. Neither A nor B   (G.1)

63. The heated mirrors will not work when the rear defogger is turned on. The rear defogger operates correctly. Technician A says the rear defogger switch is faulty. Technician B says the driver's side mirror has a faulty ground. Who is correct?
    A. A only
    B. B only
    C. Both A and B
    D. Neither A nor B   (H.1.14)

64. All of these statements about air bag system service are true **EXCEPT:**
    A. the negative battery cable should be disconnected and manufacturer's recommended waiting period completed.
    B. safety glasses and gloves should be worn when handling deployed air bags.
    C. a 12-V powered test light may be used to test continuity between the inflator module and the sensors.
    D. sensor operation may be affected if the sensor brackets are bent or twisted.   (H.2.13)

65. A cruise control with a vacuum servo and PCM operated solenoids provides erratic vehicle speed when it is engaged. Technician A says to check the vacuum hose from the intake manifold to the servo for leaks. Technician B says to check the servo diaphragm for leaks. Who is correct?
    A. A only
    B. B only
    C. Both A and B
    D. Neither A nor B   (H.2.7)

66. A vehicle with speed-sensitive wipers is being discussed. The wipers do not change speed as vehicle speed increases, but otherwise operate normally. Technician A says that the wiper motor may be defective. Technician B says that the vehicle speed sensor (VSS) may be defective. Who is right?
    A. A only
    B. B only
    C. Both A and B
    D. Neither A nor B   (G.4)

67. All of the following statements about radio antenna diagnosis with an ohmmeter are true **EXCEPT:**
    A. continuity should be present between the end of the antenna mast and the center pin on the lead-in wire.
    B. continuity should be present between the ground shell of the lead-in wire and the antenna mounting hardware.
    C. no continuity should be present between the center pin on the lead-in wire and the ground shell.
    D. continuity should be present between the end of the antenna mast and the antenna mounting hardware.   (H.2.2)

68. The windshield washer pump doesn't work. When disconnected from the vehicle and hooked to a known good power supply the motor does not operate. Which of the following should the technician do next?
    A. Test the multi-function switch.
    B. Test the pump wiring harness on the vehicle.
    C. Replace the pump.
    D. Check the hoses and nozzles for plugging.   (G.7)

69. The vehicle's antitheft system will not disarm when the key is inserted in the driver's door. Technician A tries the key in the passenger side to see if the system disarms. Technician B checks the wiring diagram of the system to identify the wires and connectors in the system. Who is right?
    A. A only
    B. B only
    C. Both A and B
    D. Neither A nor B   (H.2.10)

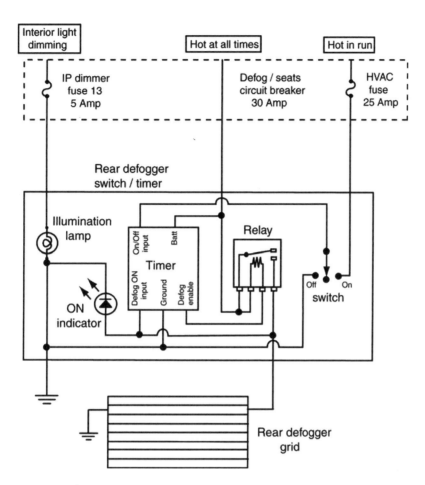

70. When the rear defogger switch is turned on, the rear defogger light is illuminated, but there is no defogger grid operation as shown in the figure. The cause of this problem could be a(n):
    A. open defogger relay winding.
    B. open circuit at the defogger relay contacts.
    C. open circuit between the switch/timer and the grid.
    D. defective defogger on/off switch.   (H.1.5)

71. An alternator with a 90 ampere rating produces 45 amperes during an output test. The alternator is driven with a V-belt and the belt has the specified tension. Technician A says the V-belt may be worn and bottomed in the pulley. Technician B says the alternator pulley may be misaligned with the crankshaft pulley. Who is correct?
    A. A only
    B. B only
    C. Both A and B
    D. Neither A nor B   (D.8)

72. None of the door locks function in the lock mode. With the lock button depressed, 12V are supplied to the lock relay. The cause of this problem could be a(n):
    A. open ground connection at the unlock relay contacts.
    B. open circuit in the LF door lock motor armature.
    C. open circuit between the master switch and the lock relay winding.
    D. failed fuse connected to the door lock switches.   (H.1.8)

73. All of the following statements about using a lab scope for diagnosis are correct **EXCEPT:**
    A. an upward movement of the trace means the voltage has increased and a downward movement means the voltage has decreased.
    B. the size and clarity of the trace is dependent on the cleanness of the connection and the component or circuit being tested.
    C. a flat waveform means the voltage is staying at that level.
    D. as the trace moves across the screen of the oscilloscope, time is represented.   (A.5)

74. The chime does not work when the driver door is opened. It works when the passenger door is open. Technician A says the problem could be the driver door switch is malfunctioning. Technician B says the problem could be the chime module. Who is right?
    A. A only
    B. B only
    C. Both A and B
    D. Neither A nor B   (F.7)

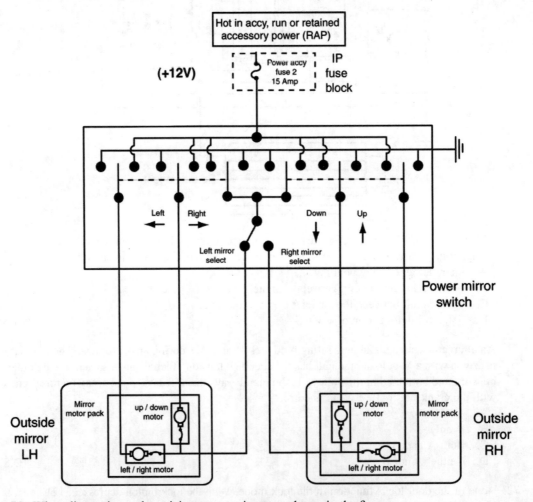

75. When diagnosing and servicing power mirrors as shown in the figure:
    A. each mirror contains a combination up/down and left/right motor.
    B. each mirror contains a separate up/down and left/right motor.
    C. the mirror motors are grounded at the mirror assemblies.
    D. when one of the switches is pressed, 9.5V are supplied to a mirror motor.   (H.1.13)

76. While discussing a radio static problem, Technician A says there might be a poor metal-to-metal connection between the hood and other body components. Technician B says the suppression coil might be defective on the instrument voltage limiter. Who is right?
    A. A only
    B. B only
    C. Both A and B
    D. Neither A nor B   (H.2.1)

77. The circuit breaker in the power seat circuit opens and closes continually without touching the seat control buttons. Technician A says something stored under the seat might be jammed in the seat tracks. Technician B says there might be a grounded wire between the circuit breaker and the seat switches. Who is right?
    A. A only
    B. B only
    C. Both A and B
    D. Neither A nor B   (H.1.3)

78. A radio has a whining noise that increases with engine speed. When the alternator field wire is disconnected, the noise stops. All of these defects may be the cause of the problem **EXCEPT** a(n):
    A. defective stator.
    B. defective diode.
    C. defective capacitor.
    D. open field winding.   (H.2.4)

79. While diagnosing slower than normal wiper operation, Technician A says the wiper linkage may be binding. Technician B says there may be excessive resistance in the power circuit. Who is right?
    A. A only
    B. B only
    C. Both A and B
    D. Neither A nor B   (G.5)

80. The charging system voltage on a vehicle is 16.2V. This condition may cause all the following problems **EXCEPT:**
    A. an overcharged battery.
    B. burned-out electrical components.
    C. electrolyte gassing in the battery.
    D. reduced headlight brilliance.   (D.5)

6

Additional Test Questions for Practice

Additional Test Questions

Please note the letter and number in parentheses following each question. They match the task in Section 4 that discusses the relevant subject matter. You may want to refer to the overview using the cross-referencing key to help with questions posing problems for you.

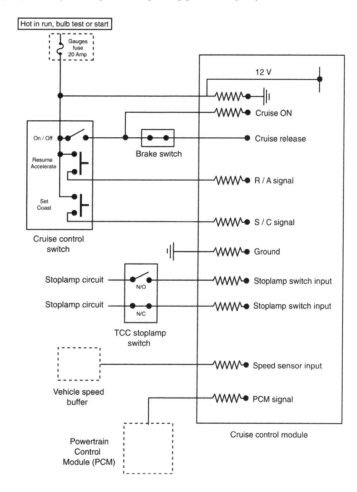

1. The cruise control is inoperative as shown in the figure. Technician A says the vehicle-speed sensor might be defective. Technician B says the 20-amp gauge fuse might be defective. Who is right?
    A. A only
    B. B only
    C. Both A and B
    D. Neither A nor B

(H.2.8)

2. The circuit breaker that protects an electric window circuit opens whenever an attempt is made to lower the window. Technician A says the internal resistance of the motor is too high. Technician B says the window regulator may be sticking. Who is right?
   A. A only
   B. B only
   C. Both A and B
   D. Neither A nor B   (H.1.1)

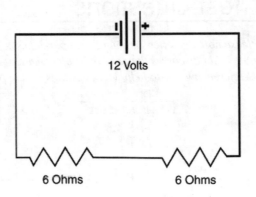

3. The total resistance in the figure is:
   A. 3 ohms.
   B. 12 ohms.
   C. 1.2 ohms.
   D. 6 ohms.   (A.3)

4. A full field test is performed on a vehicle with an internal regulator. Technician A says that if the output is within specifications, a faulty regulator may be the cause. Technician B says that an alternator should never be full-fielded for more than 30 seconds. Who is right?
   A. A only
   B. B only
   C. Both A and B
   D. Neither A nor B   (D.4)

5. Which of the following is an approved method of disarming the air bag system?
   A. Disconnect the battery.
   B. Remove the airbag fuse.
   C. Both A and B
   D. Neither A nor B   (H.2.12)

6. A voltmeter that is connected across the input and output terminals of an instrument cluster illumination lamp rheostat (dimmer control) indicates 12.6 volts with the switch in the maximum brightness position and the engine off. Which of the following statements is true?
   A. The voltage available at the lamps will be 12.6 volts.
   B. The voltage available at the lamps will be 0.0 volts.
   C. The rheostat is working normally.
   D. More information is needed in order to determine whether the lamps will operate correctly.   (E.10)

7. The brake lights will not work on a vehicle. Everything else works normally. Technician A says the problem may be a faulty hazard switch. Technician B says the problem may be a faulty brake light switch. Who is correct?
   A. A only
   B. B only
   C. Both A and B
   D. Neither A nor B   (E.2.2)

8. The horn sounds weakly anytime the horn is activated. All other systems work fine. The horn uses a relay circuit. Which of the following is the **MOST likely** problem?
   A. Poor ground on the control side of the horn relay.
   B. Faulty ground on the horn.
   C. Poor power supply to the horn switch
   D. Faulty horn switch ground   (G.2)

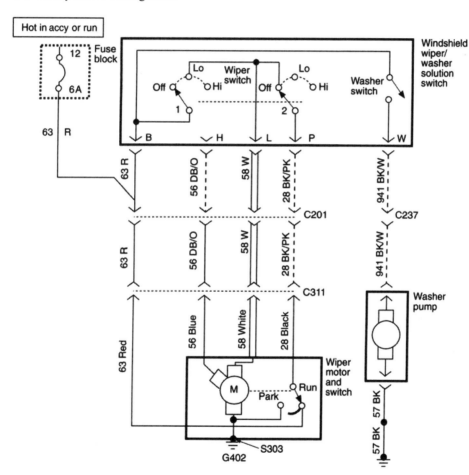

9. The low-speed position of the windshield-wiper system in the figure does not work; the washer pump works normally. Technician A says circuit 58 may be open. Technician B says circuit 63 may be open at the fuse. Who is right?
   A. A only
   B. B only
   C. Both A and B
   D. Neither A nor B   (G.3)

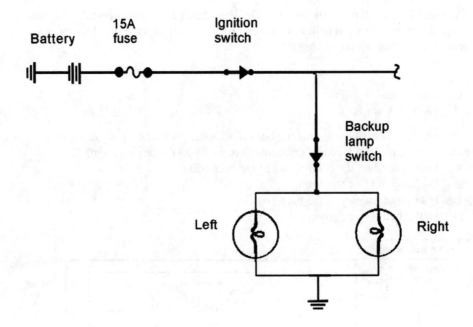

10. The back-up lights shown in the figure are inoperative. The fuse is good. With the gear selector in reverse and the ignition switch on, the voltage measured on both sides of the backup light bulbs is 0 volts. The cause of this problem could be:
    A. shorted filaments in the back-up light bulbs.
    B. an open circuit between one of the back-up lights and ground.
    C. low resistance in the back-up light switch contacts.
    D. the wire from the back-up switch to the bulbs is open.   (E.2.5)

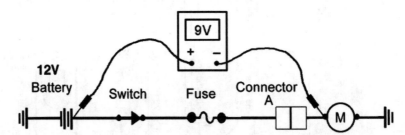

11. The voltmeter in the circuit shown reads 9 volts. Technician A says that there could be high resistance in connector A. Technician B says that there could be high resistance in the switch. Who is right?
    A. A only
    B. B only
    C. Both A and B
    D. Neither A nor B   (A.2)

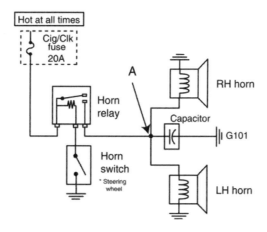

12. With the horn switch depressed, the voltage level measured at point A in the figure will be:
    A. 9V.
    B. 9.5V.
    C. 12V.
    D. 6V.   (A.2)

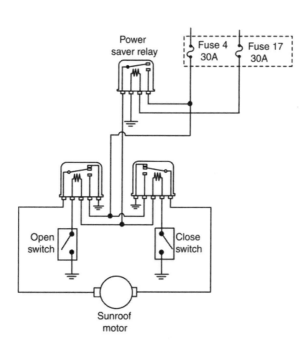

13. In the power sunroof circuit in the figure, fuse 17 fails repeatedly. Technician A says the wire between fuse 17 and the power-saver relay winding might be shorted to the ground. Technician B says the wire might be shorted to the ground between the power-saver relay winding and the ground connection. Who is right?
    A. A only
    B. B only
    C. Both A and B
    D. Neither A nor B   (H.1.11)

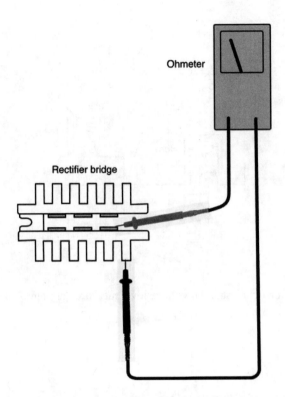

14. When testing diodes, connect the ohmmeter leads across each diode and then reverse the leads
    (see the drawing). A satisfactory diode provides:
    A. one high meter reading and one low meter reading.
    B. two meter readings of infinity.
    C. two low meter readings.
    D. a meter reading of 2 ohms and 40 ohms.   (A.10)

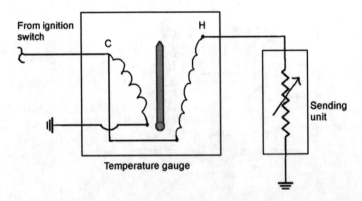

15. The temperature gauge in the figure provides a continual low reading with the engine at normal
    operating temperature. Technician A says the wire might be grounded between the gauge and
    the temperature sending unit. Technician B says the wire from the temperature gauge to the
    ground might have an open circuit. Who is right?
    A. A only
    B. B only
    C. Both A and B
    D. Neither A nor B   (F.1)

16. The cigar lighter requires repeated fuse replacements. The cause of this problem could be:
    A. high resistance in the lighter element.
    B. a high resistance connector on the cigar lighter.
    C. a shorted lighter element.
    D. a grounded wire between the lighter and the chassis.   (H.2.5)

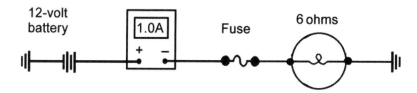

17. The current flow indicated on the ammeter shown is 1 ampere and the specified current flow is 2 amperes. The cause of this problem could be:
    A. low resistance in the light filament.
    B. high resistance at one of the light terminals.
    C. the circuit is grounded between the fuse and the light.
    D. the circuit is open between the light and ground.   (A.3)

18. The convertible top operates when the down button is pressed, but the motor stalls when the up button is pressed. The cause of this problem might be:
    A. an open circuit breaker in the convertible-top motor.
    B. an open ground wire on the convertible-top switch.
    C. a jammed linkage mechanism on the convertible top.
    D. an open ground on the convertible-top motor.   (H.1.11)

19. The RH turn signals operate normally and the LH turn signals flash rapidly. Technician A says there might be a grounded wire between the signal light switch and the RF signal light. Technician B says there might be high resistance between the LF signal light and ground. Who is right?
    A. A only
    B. B only
    C. Both A and B
    D. Neither A nor B   (E.2.3)

20. Both headlights on a vehicle have normal brilliance with the engine idling. When the engine is accelerated, both headlights become considerably brighter. The cause of this problem could be:
    A. high charging system voltage.
    B. high resistance in the headlight circuit.
    C. a shorted alternator diode.
    D. high resistance in the alternator battery wire.   (E.1.1)

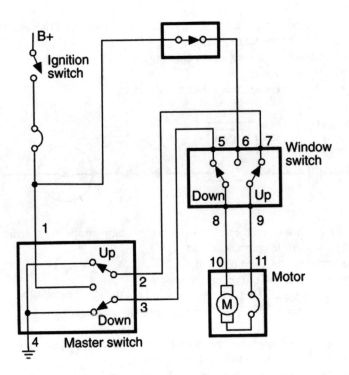

21. The power window motor in the figure is completely inoperative. With the master window switch placed in the "down" position, the following voltages are measured at each terminal:

Terminal #	Voltage
1	12V
2	0V
3	12V
4	0V
5	12V
6	12V
7	0V
8	12V
9	0V
10	12V
11	0V

Which of the following statements represents the cause of this problem?
A. The master switch is faulty.
B. The window switch is faulty.
C. The motor is faulty.
D. There is a poor ground in the circuit.

(H.1.2)

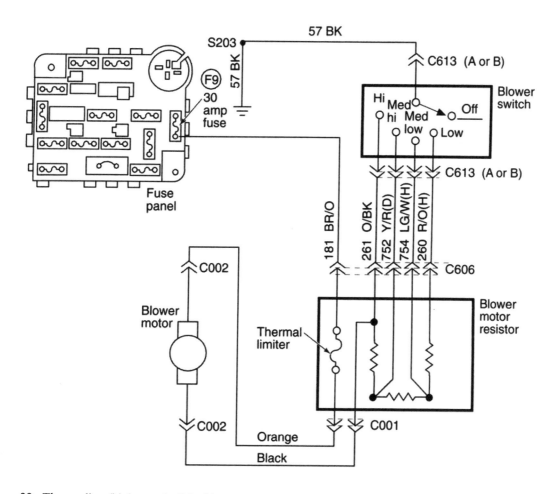

22. The medium/high speed of the blower motor circuit in the figure is inoperative; the rest of the blower speeds are fine. Technician A says circuit 752 may be open. Technician B says the middle resistor in the blower motor resistor assembly may be open. Who is right?

    A. A only
    B. B only
    C. Both A and B
    D. Neither A nor B

                                                                                (A.11)

23. Technician A says that gauge circuits that use an instrument voltage regulator (IVR) will produce inaccurate readings if the alternator is overcharging. Technician B says that a sending unit on a vehicle without an IVR can be tested by grounding the sending unit wire. Who is right?

    A. A only
    B. B only
    C. Both A and B
    D. Neither A nor B

                                                                                (F.1)

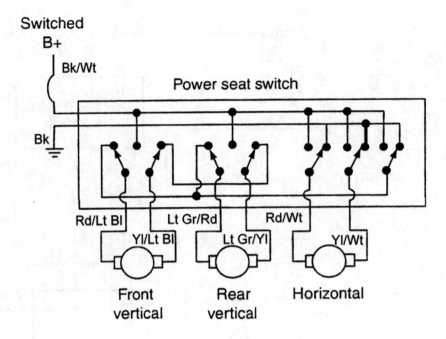

24. The rear-vertical motor in the power seat shown does not operate. Technician A says that during normal operation, there should be 12 volts at the Lt Gr/Rd wire near the motor when the motor is moving up. Technician B says that an open in the Lt Gr/YL wire could be the cause. Who is right?
    A. A only
    B. B only
    C. Both A and B
    D. Neither A nor B   (H.1.3)

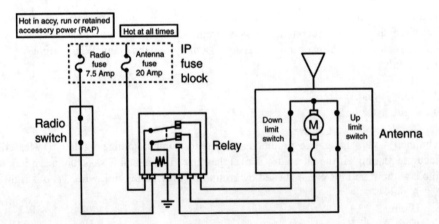

25. Referring to the figure, a customer complained the power antenna went up when the radio was turned on, but would not go back down when the car was turned off. The cause of this problem could be an open circuit in the:
    A. radio fuse.
    B. antenna fuse.
    C. open down limit switch.
    D. shorted antenna motor.   (H.2.3)

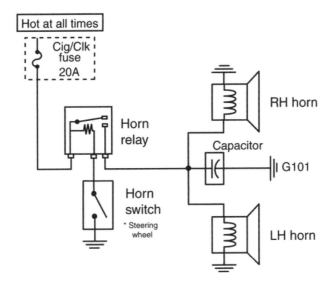

26. The horn in the diagram does not operate. Technician A says that an open winding in the horn relay may be the cause. Technician B says that a shorted horn switch may be the cause. Who is right?

    A. A only
    B. B only
    C. Both A and B
    D. Neither A nor B   (G.1)

27. Which of the following are true about light circuits?

    A. Signal and stoplights often share a common bulb filament.
    B. Many back-up lights and stoplights often share the same bulb.
    C. The taillight and stoplights share a common bulb filament.
    D. When the brake pedal is depressed, the stoplight switch is open.   (E.1.8)

28. A customer complains about repeated failure of the taillight fuse. The MOST likely cause of this problem is:

    A. a grounded stoplight wire at the back of the vehicle.
    B. an intermittent open circuit in a taillight wire.
    C. an intermittent ground in a taillight wire.
    D. a loose taillight wiring connector.   (E.1.4)

29. An alarm system will not activate when the driver door is opened. Technician A says to check the driver door-ajar switch for corrosion. Technician B says to check that the interior lights are on when the driver door is open. Who is right?

    A. A only
    B. B only
    C. Both A and B
    D. Neither A nor B   (H.2.9)

30. When cleaning, servicing, and replacing batteries, Technician A says when disconnecting battery cables, disconnect the positive cable first. Technician B says if the built-in battery hydrometer indicates black, the charging system should be tested. Who is right?
    A. A only
    B. B only
    C. Both A and B
    D. Neither A nor B   (B.4)

31. While cranking the engine, the voltmeter reading shows 0.1 volt. The cause of this reading could be:
    A. excessive resistance in the starter ground circuit.
    B. normal resistance in the positive battery cable.
    C. normal resistance in the starter ground circuit.
    D. excessive resistance in the positive battery cable.   (C.2)

32. When diagnosing and testing a rear defogger circuit,:
    A. the voltage drop from the ground side of the grid to an engine ground should not exceed one volt.
    B. a 12-V test light should be illuminated at half brilliance at any place on a grid track.
    C. if a 12-V test light is not illuminated on part of a grid track, the track is open.
    D. a 12-V test light should become brighter as it is moved to the ground side of the track.   (H.1.6)

33. Technician A says an open circuit in the instrument cluster light rheostat might cause all of the bulbs to be inoperative. Technician B says an open circuit in one of the bulbs might cause all of the bulbs to be inoperative. Who is right?
    A. A only
    B. B only
    C. Both A and B
    D. Neither A nor B   (E.1.8)

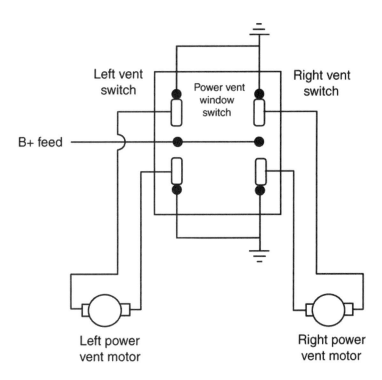

34. There is no operation from the RR power vent in the figure. The LR power vent operates normally. Technician A says there may be an open circuit between the RR power vent switch and ground. Technician B says the up/down contacts in the RR power vent switch may be open. Who is right?
    A. A only
    B. B only
    C. Both A and B
    D. Neither A nor B   (H.1.2)

35. A starting motor does not disengage properly and continues to run with the engine for a few seconds after the engine starts. Technician A says the starting motor solenoid may be the problem. Technician B says the starting motor may be misaligned with the flywheel housing. Who is correct?
    A. A only
    B. B only
    C. Both A and B
    D. Neither A nor B   (C.5)

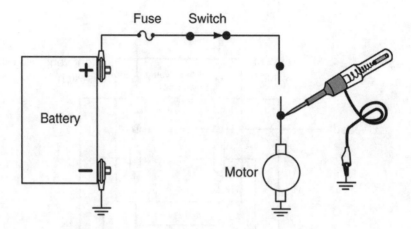

36. When the open circuit at the connector shown is repaired and the switch is closed, Technician A says the 12-V test light should be illuminated when it is connected from the motor ground brush to ground. Technician B says the 12-V test light should be illuminated at partial brilliance when it is connected from the input motor brush to ground. Who is right?
    A. A only
    B. B only
    C. Both A and B
    D. Neither A nor B   (A.1)

37. A composite headlight bulb is being replaced. Technician A says that touching the bulb with your finger may shorten its life. Technician B says that dielectric grease should be used to coat the connector terminals. Who is right?
    A. A only
    B. B only
    C. Both A and B
    D. Neither A nor B   (E.1.3)

38. A vehicle cranks slowly during starting, and current draw is high. Technician A says this could be caused by worn starter bushings. Technician B says this could be caused by high resistance in the battery ground cable. Who is right?
    A. A only
    B. B only
    C. Both A and B
    D. Neither A nor B   (C.1)

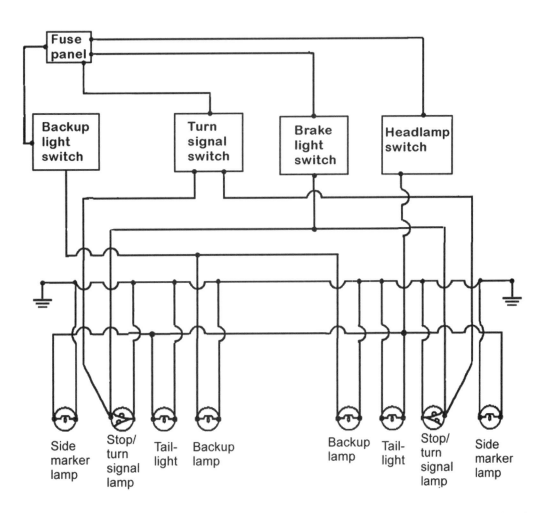

39. The rear light ground connection on the right side of the diagram shown has an open circuit. The ground connection on the left side is satisfactory. This problem could result in:
    A. inoperative LR tail and stoplights only.
    B. inoperative back-up lights.
    C. inoperative LR tail, stop, and side marker lights.
    D. no change in operation of the rear lights.   (E.1.8)

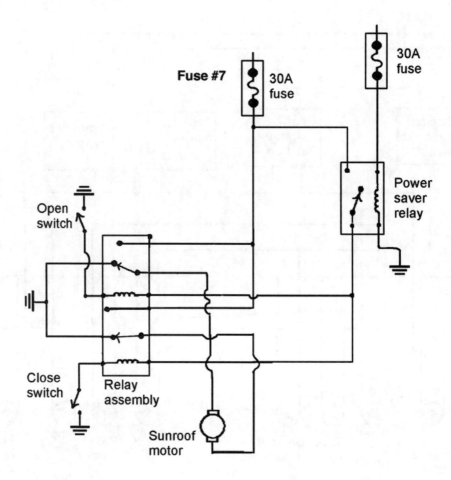

40. The sunroof in the diagram opened but failed to close. Which of the following is the MOST likely cause?
    A. A shorted close switch
    B. An open in the close relay control circuit
    C. An open in the power saver relay control circuit
    D. Fuse 7 open (blown)   (H.1.11)

41. A voltmeter connected across a starter solenoid's battery and motor terminals reads 0 volts when the ignition switch is turned to the "start" position. A distinct click is heard from the solenoid when this occurs, but the starter does not rotate. Technician A says there is excessive voltage drop in the circuit between the battery and the starter solenoid. Technician B says the starter solenoid needs to be replaced. Who is right?
    A. A only
    B. B only
    C. Both A and B
    D. Neither A nor B   (C.3)

42. An alternator has no output. Technician A says the alternator field circuit may have an open circuit. Technician B says the fuse link may be open in the alternator-to-battery wire. Who is right?
    A. A only
    B. B only
    C. Both A and B
    D. Neither A nor B   (D.1)

43. The turn signal lights operate normally, but the hazard lights are inoperative. The MOST likely cause of this problem is a(n):
    A. defective turn signal light flasher.
    B. defective hazard flasher.
    C. open circuit between the fuse and the signal light flasher.
    D. defective signal light switch.   (E.2.3)

44. A vehicle cranks slowly during starting, and current draw is low. Technician A says that low engine compression could be the cause. Technician B says that corroded battery terminals could be the cause. Who is right?
    A. A only
    B. B only
    C. Both A and B
    D. Neither A nor B   (C.1)

Appendices

Answers to the Test Questions for the Sample Test Section 5

1.	B	21.	A	41.	A	61.	C
2.	A	22.	A	42.	A	62.	A
3.	C	23.	C	43.	B	63.	D
4.	C	24.	A	44.	C	64.	C
5.	A	25.	C	45.	C	65.	C
6.	B	26.	B	46.	A	66.	B
7.	D	27.	B	47.	C	67.	D
8.	A	28.	B	48.	C	68.	C
9.	D	29.	D	49.	B	69.	C
10.	B	30.	B	50.	A	70.	C
11.	B	31.	C	51.	B	71.	A
12.	C	32.	D	52.	D	72.	A
13.	B	33.	A	53.	C	73.	B
14.	C	34.	C	54.	B	74.	A
15.	A	35.	C	55.	A	75.	B
16.	A	36.	C	56.	B	76.	C
17.	C	37.	A	57.	D	77.	B
18.	C	38.	C	58.	C	78.	D
19.	B	39.	C	59.	A	79.	C
20.	D	40.	C	60.	B	80.	D

Explanations to the Answers for the Sample Test Section 5

Question #1
Answer A is wrong. Slow engine cranking can be caused by over advanced ignition timing.
Answer B is correct. All of the answers are true except B. A shorted neutral safety switch will allow the engine to start in any gear. The neutral safety switch is used on vehicles equipped with automatic transmissions. It opens the starter control circuit when the transmission shift selector is in any position except PARK or NEUTRAL.
Answer C is wrong. Misaligned starter mounting can cause slow cranking speed.
Answer D is wrong. Slow cranking speed can be caused by a low battery state of charge.

Question #2
Answer A is correct. A shorted diode in a generator will typically cause the amperage output to be about one-third less than normal. A shorted diode may be verified by using an oscilloscope. Upon disassembly, the individual diodes can be tested with the diode check function of the multimeter.
Answer B is wrong. A broken brush would probably cause no or zero output, so answer B is wrong.
Answer C is wrong. Since the test in the question was conducted with the system full-fielded, the regulator will not have an effect on the output.
Answer D is wrong. If the capacitor in the alternator were faulty, the system's output would be erratic or there would be zero output.

Question #3
Answer A is wrong. A newspaper jammed could cause this condition.
Answer B is wrong. An open switch would stop current flow and would cause this problem.
Answer C is correct. This is another **EXCEPT**-type question. The only answer that would not affect horizontal movement is C. An open in the switch to ground circuit would affect the total operation of the seat. A jammed seat track, an open between the switch and the horizontal motor, and burned contact in the horizontal seat switch would all hinder or prevent horizontal movement. All modern six-way power seats use a reversible, permanent magnet trimotor. The motor may transfer rotation to a rack-and-pinion or to a worm-gear-drive transmission. A typical control switch consists of a four-position knob and a set of two-position switches. The four-position knob controls the forward, rearward, up, and down movement of the seat. The separate two-position switches are used to control the front tilt and rear tilt of the seat.
Answer D is wrong. Burned contacts would stop or greatly reduce current flow and would cause this problem.

Question #4
Answer A is wrong. An open circuit has no current flow, and therefore A is not correct.
Answer B is wrong. A short would cause higher than normal current flow. Ohm's Law is the basic law of electricity. It states that the amount of current in an electric circuit is inversely proportional to the resistance of the circuit, and it is directly proportional to the voltage in the circuit. For example, if the resistance decreases and the voltage remains constant, the amperage will increase. If the resistance stays the same and the voltage increases, the amperage will also increase.
Answer C is correct. Lower amperage is always caused by higher than normal resistance somewhere in the circuit.
Answer D is wrong. A blown fuse would cause NO current flow.

Question #5
Answer A is correct. Only Technician A is correct. The ignition key must be on.
Answer B is wrong. The engine does not need to be running to communicate with the BCM.
Answer C is wrong. Only Technician A is correct.
Answer D is wrong. Only Technician A is correct.

Question #6
Answer A is wrong. If the open was in the switch, the bulb would illuminate when the jumper wire bypassed the switch.
Answer B is correct. With the jumper wire connected from the ground side of the bulb and a good ground, the normal ground circuit is bypassed. This caused the light to illuminate. **Closed circuit** is an electrical term that means that there are no breaks in the path and the current will flow. **Open circuit** is used to mean that current flow is stopped. By opening the circuit, the path for electron flow is broken.
Answer C is wrong. There is no circuit breaker in this circuit.
Answer D is wrong. If the fuse was burned out the bulb would illuminate when the jumper wire bypassed it.

Question #7
Answer A is wrong. If all else were normal, the voltage drop across the lamp would be 12 volts, regardless of the resistance of the light.
Answer B is wrong. If the circuit were grounded, as suggested by Technician B, there would be 0 volts across the light and the fuse would open due to increased current flow.
Answer C is wrong. Neither Technician is correct.
Answer D is correct. Neither Technician is correct. This is a 12-volt circuit. If the voltage drop across the lamp is 9 volts (as shown), that means 3 volts are being dropped either before the lamp or at the ground of the lamp.

Question #8
Answer A is correct. In order to measure current flow through the tester, the switch must be closed or turned on. While doing this test, one of the potential problems is hurrying through it. Time must be allowed for computer systems to go into sleep mode.
Answer B is incorrect. Answer B allows for too much battery drain; on most vehicles the maximum allowable drain is 50 milliamps.
Answer C is incorrect. The actual drain on the battery does not show up immediately, as suggested by answer C. It may not be shown on the meter until several minutes have passed because various computer and security system features must be allowed to "time out." The time for this to happen must be determined using the service manual.
Answer D is incorrect. Since an open door switches dome and courtesy lights on, there will be a drain on the battery when the door is open. This would prevent measuring true parasitic load, as the current drain from the door would increase the measured drain.

Question #9
Answer A is wrong. A good circuit breaker is closed until high current flows through it, and then it opens. Technician A suggests a good circuit breaker is one that is open not closed and is wrong. A good circuit breaker will have zero resistance.
Answer B is wrong. Technician B knows that an ohmmeter runs current through the component being tested in order to measure its resistance. However, the amount of current provided by the meter is very low, too low to cause the circuit breaker to trip or open.
Answer C is wrong. Neither Technician is correct.
Answer D is correct. Neither Technician is correct. An infinite reading on an ohmmeter indicates there is an open. A circuit breaker is typically abbreviated c.b. in a fuse chart of a service manual.

Question #10
Answer A is wrong. When checking a fuse with a test light, the test light will illuminate when it is touched to both sides of the fuse. If the fuse is bad, the light will illuminate at only one side. In this case, the test light never comes on, so we know nothing about the fuse and A is wrong.
Answer B is correct. Only Technician B is correct. Technician B is correct in stating there is no power to the fuse since the test light did not illuminate at either side. There are three basic types of fuses: glass or ceramic fuses, blade-type fuses, and bullet or cartridge fuses. Glass and ceramic fuses are found mostly on older vehicles. Sometimes, however, you can find them in a special holder connected in series with a circuit. Glass fuses are small glass cylinders with metal caps. The metal strip connects the two caps. The rating of the fuse is normally marked on one of the caps.
Answer C is wrong. Only Technician B is correct.
Answer D is wrong. Only Technician B is correct.

Question #11
Answer A is wrong. This is a diode symbol.
Answer B is correct. If you had problems, refer to the Overview Section of this guide and study the symbols shown there. Also look in a good textbook or service manual for more electrical schematic symbols.
Answer C is wrong. This is a diode symbol.
Answer D is wrong. This is a diode symbol.

Question #12
Answer A is incorrect. Technician B is also correct.
Answer B is incorrect. Technician A is also correct.
Answer C is correct. Both Technicians are correct. A battery's state of charge can be measured two ways: with a voltmeter and with a hydrometer. The voltmeter is connected across the battery and will measure at least 12.6 volts if the battery is fully charged. The hydrometer measures the specific gravity of the electrolyte. When all the cell readings exceed 1.265, the battery is fully charged. When checking the state of charge with a voltmeter, this check is sometimes referred to as OCV (open circuit voltage).
Answer D is wrong Both Technicians are correct.

Question #13
Answer A is wrong. While conducting a battery load test, the test load is determined by multiplying the cold-cranking amperage rating by 0.5 (50 percent), by multiplying the amp-hour rating by three or by referring to the specifications given by the manufacturer. Technician A is using the wrong formula to determine the test load.
Answer B is correct. Only Technician B is correct. Technician B is looking at the battery's voltage during the load test. A good battery will maintain 9.6 volts under load. The **capacity test** provides a realistic determination of the battery's condition. For this test to be accurate, the battery must pass the state of charge or open circuit voltage test. If it does not, recharge the battery and test it again.
Answer C is wrong. Only Technician B is correct.
Answer D is wrong. Only Technician B is correct.

Question #14
Answer A is wrong. This will not damage all the computers.
Answer B is wrong. The engine will restart when the battery cable is reconnected.
Answer C is correct. Correctly disconnecting the battery on late-model vehicles will not cause any problems to the computer or to the vehicle itself. The only thing that will happen is that items held in memory by the computers and accessories may disappear. The other answers are invalid as long as the battery is disconnected and reconnected properly. The negative cable should be the first cable to be disconnected and the last one to be reconnected.
Answer D is wrong. Voltage surges will not be created as long as the disconnect is done properly.

Question #15
Answer A is correct. Only Technician A is correct. A short always causes higher-than-normal current flow, except when it is after all loads in the circuit. So Technician A is correct. Whenever there is resistance and current flow in a circuit, voltage is dropped across the resistance.
Answer B is wrong. Technician B is incorrect in saying that higher resistance in a winding will cause a higher-than-normal voltage drop across the winding. When there is a single load in a circuit, all the voltage will be dropped across that load regardless of its resistance.
Answer C is wrong. Only Technician A is correct.
Answer D is wrong. Only Technician A is correct.

Question #16

Answer A is correct. Only Technician A is correct. It is natural for dirt and grease to collect on the top of a battery. If it is allowed to accumulate, the dirt and grease can form a conductive path between the battery terminals, causing battery drain. Also, normal battery gassing will deposit sulfuric acid as the vapors condense. Over a period of time, the sulfuric acid will corrode the battery terminals, cable clamps, and hold down fixtures. The battery should be cleaned using a solution of baking soda and water.

Answer B is incorrect. Normal gassing of the battery will cause this accumulation.

Answer C is incorrect. Only Technician A is correct.

Answer D is incorrect. Only Technician A is correct.

Question #17

Answer A is wrong. Technician B is also correct.

Answer B is wrong. Technician A is also correct.

Answer C is correct. Both Technicians are correct. Although fast charging a battery is not the preferred way to charge, sometimes it is necessary. Fast charging requires that the battery be monitored at all times. Never fast charge a battery for longer than two hours. Also, never allow the voltage of a 12-volt battery to exceed 15.5 volts. Nor should you ever allow the temperature to rise above 125° F. Fast charging the battery will bring the state of charge up high enough to crank the engine. However, fast charging is unable to recharge the battery as effectively as slow charging. Fast charging the battery converts only the lead sulfate on the outside of the plates. The conversion does not go through the plates. After the battery has been fast charged to a point that it will crank the engine, it should then be slow charged to a full state.

Answer D is wrong. Both Technicians are correct.

Question #18

Answer A is wrong. When servicing the battery and electrical system in an air-bag equipped vehicle, the negative battery cable should be disconnected first.

Answer B is wrong. The positive battery cable should be reconnected first.

Answer C is correct. After battery cables are disconnected, the Technician must wait for the time-specified time period before servicing the electrical system.

Answer D is wrong. A 12V source should be connected to the cigarette lighter socket before disconnecting the battery.

Question #19

Answer A is wrong. .Whenever jump-starting a vehicle, it is recommended that the accessories in both vehicles be turned off. When jump-starting a vehicle, make sure the vehicles are not touching each other. Turn off the accessories and make sure the parking brake is set in both vehicles.

Answer B is correct Only Technician B is correct. Connect the positive booster cables first. Then connect one end of the negative cable to the booster battery's negative terminal and the other end to an engine ground on the vehicle being boosted. Once the engine of the boosted vehicle starts, disconnect the negative cable from the engine ground. Then disconnect the rest of the cables.

Answer C is incorrect. Only Technician B is correct.

Answer D is incorrect. Only Technician B is correct.

Question #20

Answer A is wrong. Never connect an ohmmeter to a powered circuit because meter damage can occur.

Answer B is wrong. An open circuit breaker, fuse, or fuse link causes an infinite ohmmeter reading.

Answer C is wrong. Neither Technician is correct.

Answer D is correct. Neither Technician is correct. When an ohmmeter is connected to a fusible link, the meter should read zero Ohms if the component is working properly. A vehicle may have one or several fusible links to provide protection for the main power wires before they are divided into smaller circuits at the fuse box. The fusible links are usually located at a main connection near the battery or starter solenoid. The current capacity of a fusible link is determined by its size. A fusible link is usually four wire sizes smaller (four numbers larger) than the circuit it protects — the smaller the wire, the larger its number. A circuit that uses 14- gauge wire would require an 18-gauge fusible link for protection.

Question #21

Answer A is correct. Keep in mind that the slower a motor turns, the more amperage it will draw. Worn bushings will cause the motor to turn slower than normal and will cause high current draw. Higher-than-specified current draw test results indicate an impedance to rotation of the starter motor. This includes worn bushings, a mechanical blockage, and excessively advanced ignition timing. Also, higher-than-specified current draw can be the result of shorted windings in the starter motor. A lower-than-normal current draw test results indicate excessive voltage drop in the circuit, faulty relay or solenoid, or worn brushes.

Answer B is wrong. High resistance always causes lower than normal current draw.

Answer C is wrong. High resistance causes decreased current draw.

Answer D is wrong. Burned terminals would cause high resistance.

Question #22

Answer A is correct. Only Technician A is correct. A short to ground always increases circuit current flow if the short is on the power side of the circuit. The increase in current can cause the circuit's fuse to blow, just as Technician A said. To confirm that the circuit has a ground before the load, remove the fuse and connect a test light in series across the fuse connections. If the lamp lights, the circuit has a short to ground. It is difficult to test for shorts to ground with a test light or voltmeter because the fuse blows before any testing can be conducted. To prevent this, connect a cycling circuit breaker that is fitted with alligator clips across the fuse holder. The circuit breaker will continue to cycle open and closed, allowing the technician to test for voltage.

Answer B is wrong. The short would affect the entire circuit, not just the side it is in. Therefore, Technician B is wrong.

Answer C is wrong. Only Technician A is correct.

Answer D is wrong. Only Technician A is correct.

Question #23

Answer A is wrong. The positive battery cable is not between the meter leads.

Answer B is wrong. The starter solenoid windings are not between the test leads.

Answer C is correct. When a voltmeter is connected across something, it displays the voltage difference between the two points. In the diagram, the voltmeter is shown connected from the negative battery post to the ground at the starter. This connection will measure the voltage drop only across the starter's ground circuit. To measure the voltage drop across the positive battery cable, the meter should be connected at the positive battery post and the starter end of the positive cable. To measure the voltage drop across the solenoid windings, the meter must be connected across the windings. The same is true for the solenoid disc and terminals. Remember, to perform this test, the starter motor must be cranking.

Answer D is wrong. The disc and terminals are not between the meter leads.

Question #24

Answer A is correct. Only Technician A is correct. A short to ground may cause excessive current flow in the light circuit. This excessive current flow may cause the circuit breaker to open the circuit and shut off the lights.

Answer B is wrong. If the charging system voltage is high enough to open the headlight circuit breaker, many other problems would occur such as damaged electronic components and an overcharged battery.

Answer C is wrong. Only Technician A is correct.

Answer D is wrong. Only Technician A is correct.

Question #25

Answer A is wrong. A corroded negative battery could be the cause.

Answer B is wrong. A loose ground connection could be the cause.

Answer C is correct. The measured voltage drop in this question is well beyond any specification and the problem needs to be fixed. All of the answers are possible locations for the problem, except answer C. High resistance in the starter field windings would not affect the measured voltage drop on the ground side of the starter, although it would affect starter operation. A slow crank or no-crank complaint can be caused by several potential trouble spots in the circuit. Excessive voltage drops in these areas will cause the starter motor to operate slower than required to start the engine. The speed that the starter motor rotates the engine is important to engine starting. If the speed is too slow, compression is lost and the air/fuel mixture draw is impeded. Most manufacturers require a speed of 100 to 250 rpm during engine cranking.

Answer D is incorrect. An improperly grounded starter could cause this problem.

Question #26

Answer A is wrong. An ohmmeter check is being taken across the solenoid S terminal and the solenoid case. The meter reads infinite which means there is no continuity between the two points. Answer A states the pull-in coil is shorted to ground. If it were, the reading would show very low resistance. Also, the ohmmeter leads must be connected across the solenoid terminal and the field coil terminal to test the pull-in winding.

Answer B is correct. Answer B cites an open in the hold-in winding. An infinite reading certainly indicates an open. To test the hold-in winding, the ohmmeter leads should be connected from the solenoid S terminal to the ground or solenoid case.

Answer C is wrong. Answer C is incorrect because the hold-in winding is being tested.

Answer D is wrong. Answer D is incorrect because the hold-in winding is open, not shorted. When the ignition switch is placed in the START position, voltage is applied to the S terminal of the solenoid. The hold-in winding has its own ground to the case of the solenoid. The pull-in winding's ground is through the starter motor. Current will flow through both windings to produce a strong magnetic field. When the plunger is moved into contact with the main battery and motor terminals, the pull-in winding is de-energized. The pull-in winding is not energized because the contact places battery voltage on both sides of the coil. The current that was directed through the pull-in winding is now sent to the motor. Because the contact disc does not close the circuit from the battery to the starter motor until the plunger has moved the shift lever, the pinion gear is in full mesh with the flywheel before the armature starts to rotate.

Question #27

Answer A is wrong. The condition is that the starter will not crank the engine. A jumper wire is used to bypass the starter control circuit. When this is done, the starter is able to crank the engine. A says this indicates the solenoid is bad. This is wrong. If the solenoid were bad, it would not work once the control circuit was bypassed.

Answer B is correct. Only Technician B is correct. A bad ignition switch would cause this problem, and it is being bypassed with the jumper wire. Refer to the manufacturer's manuals for correct procedures for disconnecting a battery. Some vehicles with onboard computers must be supplied with an auxiliary power source. Be sure the vehicle is properly positioned on the hoist or on safety jack stands. Before performing any cranking test, be sure the vehicle is in park or neutral and the parking brakes are applied. Follow manufacturer's directions for disabling the ignition system. Be sure the test leads are clear of any moving engine components. Never clean any electrical components in solvent or gasoline. Clean with compressed air, denatured alcohol, or wipe with clean rags only.

Answer C is wrong. Only Technician B is correct.

Answer D is wrong. Only Technician B is correct.

Question #28

Answer A is wrong. Although Technician A is correct by wanting to verify that the engine is or is not seized by rotating it with a large wrench, this should not be done first.

Answer B is correct. Only Technician B is correct. The first thing to do when facing a no-crank situation is to check the battery. The battery condition is quickly checked and is the most common cause of this type of customer concern.

Answer C is wrong. Only Technician B is correct.

Answer D is wrong. Only Technician B is correct.

Question #29
Answer A is wrong. Technician A is wrong. Typically a loose or glazed belt will cause decreased output at high speeds. It will not cause an overcharge condition.
Answer B is wrong. Technician B is also wrong. If the grounded field wire did full field the alternator, the output should be high.
Answer C is wrong. Neither Technician is correct.
Answer D is correct. Neither Technician is correct.

Question #30
Answer A is wrong. What we are looking at here is a 1.4-volt drop between the generator output terminal and the battery. If there were excessive resistance on the ground side of the alternator, the output at the generator would be less. If the ground for the alternator were bad, the ground at the case would be bad and the measured output lower.
Answer B is correct. Only Technician B is correct. Excessive resistance in the positive side of the circuit could only cause the voltage drop being witnessed here. To best perform this test, the alternator should be under load, thereby causing more current to flow.
Answer C is wrong. Only Technician B is correct.
Answer D is wrong. Only Technician B is correct.

Question #31
Answer A is wrong. Technician B is also correct.
Answer B is wrong. Technician A is also correct.
Answer C is correct. Both Technicians are correct. Fusible links are circuit protection devices that are connected in series within a circuit. A is correct in saying a good fusible link will have voltage at both ends of it. B is also right. Fusible links should never be longer than nine inches. Links longer than this will decrease its protection ability. Not all fusible link open circuits are detectable by visual inspection only. Test for battery voltage on both sides of the fusible link to confirm its condition. If the fusible link must be replaced, it is cut out of the circuit and a new fusible link is crimped or soldered into place. There are two types of insulation used on fusible links: Hypalon and Silicone/GXL. Hypalon can be used to replace either type of link. However, do not use Silicone/GXL to replace Hypalon. To identify the type of insulation, cut the blown link's insulation back. The insulation of the Hypalon link is a solid color all the way through. The insulation of Silicon/GXL will have a white inner core.
Answer D is wrong. Both Technicians are correct.

Question #32
Answer A is wrong. Technician A describes a probable cause for the locks not locking.
Answer B is wrong. Technician B describes a probable cause for the locks not unlocking.
Answer C is wrong. Neither Technician is correct.
Answer D is correct. Neither Technician is correct. The key to answering this question is to note that the unlock and lock functions are separate. There are two controlling circuits: one for lock and one for unlock. The problem in this question is that the locks do not work, not that they do not lock or do not unlock. So neither is correct. The problem lies in the circuit where it is common. Electric power locks use either a solenoid or a permanent magnet (PM) reversible motor. Due to the high current demands of solenoids, most modern vehicles use PM motors. Depending on circuit design, the system may incorporate a relay. The relay has two coils and two sets of contacts to control current direction. In this system, the door lock switch energizes one of the door lock relay coils to send battery voltage to the motor.

Question #33

Answer A is correct. Only Technician A is correct. Most headlight circuits are equipped with self- or automatic-reset circuit breakers. This is what is happening here. The headlights turn on and off at will because the breaker is opening and closing. The tripping of the breaker is probably caused by an intermittent short.

Answer B is wrong. Technician B is wrong. A high charging rate would not cause the breaker to trip; nor would the lights turn on and off, they would be brighter or they would be damaged by high system voltage. Headlights that flash on and off as the vehicle goes over road irregularities indicate a loose connection. Headlights that flash on and off at a constant rate indicate that the circuit breaker is being tripped. There is an overload in the circuit that must be traced and repaired, or it is a failed circuit breaker.

Answer C is wrong. Only Technician A is correct.

Answer D is wrong. Only Technician A is correct.

Question #34

Answer A is wrong. A defective headlamp door is not LEAST likely problem.

Answer B is wrong. Loose or broken vacuum lines are not the LEAST likely problem.

Answer C is correct. The LEAST likely cause of the problem in this question is C, a poor headlamp ground. This problem would affect the brilliance of the headlight, not the performance of the retractable door. The other answers or statements are correct. Electrically operated systems generally incorporate a headlamp control module, which provides power to the headlamp door motors in response to signals received from the headlamp circuit. The movement of the door is halted once in position by a limit switch. Some headlamp door retractors are vacuum operated. Any condition that would cause low vacuum will slow or prevent headlamp door operation. The door could also be stuck due to interference or lack of lubrication.

Answer D is wrong. A faulty limit switch is not the LEAST likely problem.

Question #35

Answer A is wrong. You should handle the bulb by the base only, never touch the glass.

Answer B is wrong. You should never drop or scratch the bulb; it will shorten the bulb's life.

Answer C is correct. When replacing a halogen light bulb, always turn off the headlights and allow the bulbs to cool. Keep moisture away from the bulb, and handle the bulb by its base. Do not scratch or drop the bulb. Coat the terminals with dielectric grease to minimize corrosion. To prevent early failure, do not energize the bulb unless it is installed into the socket.

Answer D is wrong. You should keep moisture away from the bulb.

Question #36

Answer A is wrong. The taillights would not work if fuse 5 were blown.

Answer B is wrong. The stoplights are not affected by fuse 5.

Answer C is correct. A look at the schematic for the question should lead you to the answer to this question. The instrument cluster or panel lights would be affected. Fuse 5 is the protection device for that circuit.

Answer D is wrong. The low beams would still work if fuse 5 were blown because they are in the circuit with fuse 4. Something after the fuse is causing too much current flow, possibly the switch, wiring, or instrument cluster.

Question #37

Answer A is correct. Only Technician A is correct. The air bag warning light is designed to illuminate whenever the system detects a fault in the system and for a few seconds after the engine is started while the module performs system checks. The light does not come on when the bag deploys. The driver and passenger will notice if the bag deploys and there is no need for a warning light. Before an air bag system is diagnosed, a system check is performed to avoid diagnostic errors. Always consult the manufacturer's specific information because the diagnostic system check may vary depending on the vehicle. The diagnostic system check involves observing the air bag warning light to determine if it is operating normally.

Answer B is wrong. The air bag will not deploy because the light is on.

Answer C is wrong. Only Technician A is correct.

Answer D is wrong. Only Technician A is correct.

Question #38

Answer A is wrong. Technician B is also correct.

Answer B is wrong. Technician A is also correct.

Answer C is correct. Both Technicians are correct. Technician A is correct. Some headlight door motors have a manual knob on the headlight door motors. If the motors do not open the doors, this knob may be rotated to lift the doors. Technician B is also correct. When the driver turns the headlights on, voltage is supplied from the battery through the headlight switch to the headlight door module. In response to this signal, the headlight door module supplies voltage to both headlight door motors. If the headlights are shut off, the module reverses the motor action to close the doors. A vehicle equipped with a concealed headlight system hides the lamps behind doors when the headlights are turned off. When the headlight switch is turned to the HEADLIGHT position, the headlight doors open. The headlight doors can be controlled by either electric motors or by vacuum. In vacuum-controlled systems, a vacuum distribution valve controls the direction of vacuum to various vacuum motors or to vent. Electrically controlled systems use either a torsion bar to open both headlight doors from a single motor or a separate motor for each headlight door.

Answer D is wrong. Both Technicians are correct.

Question #39

Answer A is wrong. Technician B is also correct.

Answer B is wrong. Technician A is also correct.

Answer C is correct. Both Technicians are correct. Daytime running lights (DRL) are typically part of the vehicle's high-beam circuit. The control circuit is connected directly to the vehicle's ignition switch, so the lights are turned on whenever the vehicle is running. The circuit is equipped with a module that reduces battery voltage to approximately 6 volts. This voltage reduction allows the high beams to burn with less intensity and prolongs the life of the bulbs. When the headlight switch is moved to the ON position, the module is deactivated and the lights work normally. Both technicians are correct. Troubleshooting these systems should begin by identifying whether the problem is in the DRL system or the headlight system. If the problem is in the headlight system, service to the circuit and lamps is conducted in the same way as for vehicles that are not equipped with DRL. If the problem is in the DRL system and the headlights work normally, only that part of the circuit that is unique to the DRL can be the problem.

Answer D is wrong. Both Technicians are correct.

Question #40

Answer A is wrong. Technician B is also correct.

Answer B is wrong. Technician A is also correct.

Answer C is correct. Both Technicians are correct. If the tone generator does not work at all, the sounding device is bad or there is an open in the circuit. If the generator works constantly, the existence of the situation the device was designed to warn the driver about or the controlling circuit is shorted and the switch or sender bypassed. If the audible warning is heard intermittently, the MOST likely cause is a loose wire or connector. Technician B says the tone generator should be checked for operation by running it through the prescribed self-test mode. This is also correct. Many of the warning systems on today's vehicles are triggered by a PCM or BCM. Always refer to the testing methods recommended by the manufacturer when testing these systems. The buzzer may have a timer circuit to control how long it is activated. If the buzzer does not shut off, replace the timer with a known good unit. If the buzzer does not have a timer circuit and remains active after the indicated problem is corrected (headlights on with the door open, for example), check for a short to ground on the switch side of the buzzer and for a faulty switch. Most door ajar, key warning, and courtesy light switches are normally open (NO) switches. If the buzzer stops sounding when the switch is unplugged, the switch is probably stuck in the closed position. If the buzzer does not sound, remove it and apply 12 volts across the terminals. If the buzzer sounds, it is good and the problem is in the circuit.

Answer D is wrong. Both Technicians are correct.

Question #41

Answer A is correct. Only Technician A is correct. When clicking dim taillight problems, determine if only one or all of the taillights are dim. If all of the taillights are dim, then there is a high resistance between the headlight switch and the common point to all the taillights. If only one taillight is dim, then there is a bad connection between the common point and the individual taillight. Most likely the taillight or the surrounding metal has corroded at the point where the taillight is grounded.

Answer B is wrong. Technician B's suggestion would affect all of the taillights. The best test for this type of condition is a voltage drop test. The portion of the circuit causing the voltage drop can easily be isolated with this test.

Answer C is wrong. Only Technician A is correction.

Answer D is wrong. Only Technician A is correction.

Question #42

Answer A is correct. Only Technician A is correct. A seized pump will not operate and there would be no fluid flow from the nozzle.

Answer B is wrong. The nozzles being plugged would cause no fluid flow, however, the pump would still operate.

Answer C is wrong. Only Technician A is correct.

Answer D is wrong. Only Technician A is correct.

Question #43

Answer A is wrong. Since the grounds are redundant the LR taillight, stoplight, and side marker lights would continue to work.

Answer B is correct. Upon examination, you will notice the grounds for the lights are redundant. So if one of the grounds is bad, the lights will still function normally. The other answers assume the loss of a ground for the circuit. If the right ground becomes overloaded, all the rear lights will be affected.

Answer C is wrong. Since the grounds are redundant the LR taillight, stoplights would continue to operate.

Answer D is wrong. Since the grounds are redundant, the back up lights would continue to operate.

Question #44

Answer A is wrong. Technician B is also correct.

Answer B is wrong. Technician A is also correct.

Answer C is correct. Both Technicians are correct. The possible causes for low-speed only operation are a faulty wiper switch, worn brushes, poor circuit connections, or an open in the control circuit. Many motors are equipped with three brushes riding on the armature. Two are located directly opposite of each other and are used for all speeds. The third brush is used for high-speed operation only. Most windshield wiper motors use permanent magnet fields. Motor speed is controlled by the placement of the brushes on the commutator. Three brushes are used: common, high speed, and low speed. The common brush carries current whenever the motor is operating. The low-speed and high-speed brushes are placed in different locations, based on motor design. The placement of the brushes determines how many armature windings are connected in the circuit. There are fewer armature windings connected between the common and high-speed brushes. When battery voltage is applied to fewer windings there is less magnetism in the armature and a lower counterelectromotive force (CEMF). With less CEMF in the armature, there is greater armature current. The greater armature current results in higher speeds.

Answer D is wrong. Both Technicians are correct.

Question #45
Answer A is wrong. Technician B is also correct.
Answer B is wrong. Technician A is also correct.
Answer C is correct. Both Technicians are correct. Instrument panel or cluster light circuits are activated when the park lights are turned on. A separate rheostat is used to control the brightness of the lamps. Power typically is provided to the lamps from the headlight switch through the rheostat. Technician A is correct. The printed circuit mounted to the back of the instrument cluster contains the electrical pathways for the instruments and the lamps. If a printed circuit board is damaged or has a fault, it must be replaced and cannot be repaired. Technician B is also correct. Be careful when testing the printed circuit. Do not touch the circuit paths with your fingers. Do not scratch the lamination with the test probes. Doing so may destroy a good circuit board.
Answer D is wrong. Both Technicians are correct.

Question #46
Answer A is correct. With an open circuit there is no current flow; therefore, there are no voltage drops. Full battery voltage would be available in the circuit until the point of the open. In the figure used for this question, the test light is probing the circuit before the open and would illuminate with full power. Any point in the circuit after the open would have no voltage; therefore, the test light would not illuminate. A test light is limited because it does not display how much voltage is at the point of the circuit being tested. However, by understanding the effects of voltage drop, the technician will be able to interpret the brightness of the test light and relate the results to the expectations of a good circuit. If the lamp is connected after a voltage drop, the lamp will light dimly. Connecting the test light before the voltage drop should light the lamp brightly. The light should not illuminate at all if it is probing for voltage after the last resistance.
Answer B is wrong. The test light should illuminate.
Answer C is wrong. If the test light were connected to the ground it should not illuminate.
Answer D is wrong. The other side of the connecter should not have voltage.

Question #47
Answer A is wrong. Answer A relates to high resistance in wire DB 180G RD. This wire belongs to the left signal circuit and would not affect the right side. It would cause dim and/or slow operation of the left side signals.
Answer B is wrong. The problem in answer B is a short to ground in the same circuit. Again, only the left side would be affected.
Answer C is correct. High resistance in wire D7 18BR RD (answer C) would affect the operation of the right side.
Answer D is wrong. Answer D is also a high resistance problem, but the chosen wire is common to all of the turn signals and therefore would not have an effect on just one side. To isolate which part of D7 18BR RD is at fault, a voltage drop test should be used.

Question #48
Answer A is wrong. Technician B is also correct.
Answer B is wrong. Technician A is also correct.
Answer C is correct. Both technicians are correct with their statements concerning turn signals. The multifunction switch can be checked with a pin-out chart and a DMM. Also a burned out bulb can cause many turn signal problems.
Answer D is wrong. Both technicians are correct.

Question #49
Answer A is wrong. Answer A would blow the fuse.
Answer B is correct. The backup lights do not work at all. This is a good indication of an open circuit. The presence of 12 volts at both sides of the bulb indicates the open is after the bulbs. The only thing after the bulbs is the ground. That must be where the open is. In fact, an open here explains why 12 volts was measured on both sides of the bulb: the circuit is incomplete and has no ground. When there is no ground, there is no current flow. Without current there will be no voltage drop across the bulbs.
Answer C is wrong. Answer C would result in less than 12 volts on the positive side of the bulb.
Answer D is wrong. Answer D would cause the lights to be on constantly.

Question #50

Answer A is correct. The MOST likely cause for an inoperative sunroof listed in this question is answer A. If there were an open in the power saver relay winding, the relay would never energize and the power to both the open and close relay windings would not be present. The sunroof would not move.

Answer B is wrong. Answer B suggests an open in close relay winding, this would effect the closing only.

Answer C is wrong. Answer C suggests an open in close relay; this would effect closing only.

Answer D is wrong. An open in the close relay and switch would affect only the close mode of operation. If the sunroof does not operate, begin by testing the fuses. Use a test light or voltmeter to test for voltage at both sides of the fuses. If voltage is present at both sides, then the fuses are good. If there is voltage into the fuses but not out of them, the fuses are bad. If there is no voltage into the fuses, then there is an open in the feed from the battery.

Question #51

Answer A is wrong. A DMM can be used to test a trailer wiring connector. However the meter will not load the circuit and may not be effective in finding a poor connection.

Answer B is correct. When testing a trailer wiring connector, the testlight will load the circuit. Most failures in this circuit deal with higher than normal resistance, which only becomes evident when current flow is near its normal level.

Answer C is wrong. Although an ohmmeter is effective at measuring resistance, in this circuit the resistance will not become evident until there is sufficient current flow. The ohmmeter is not capable of requiring the desired current flow.

Answer D is wrong. An analog voltmeter will not require sufficient current flow to diagnose this type of faulty electrical circuit.

Question #52

Answer A is wrong. Neither Technician is correct.

Answer B is wrong. Neither Technician is correct.

Answer C is wrong. Neither Technician is correct.

Answer D is correct. Neither Technician is correct. If a remote keyless entry system does not operate unless the person holding the remote control is right beside the vehicle, the battery in the remote control is likely weak. One of the first diagnostic steps would be to test the battery in the remote control.

Question #53

Answer A is wrong. High resistance in the sending unit ground wire would cause the gauge to read too low at all times. Therefore answer A is wrong.

Answer B is wrong. The same would result from the presence of high resistance between the IVR and the gauge, so B is also wrong.

Answer C is correct. The gauge in the figure for this question is a thermal-electric gauge. These gauges contain a bimetallic strip surrounded by a heating coil. The pivoted gauge pointer is connected to the bimetallic strip. The sending unit contains a variable resistor. In a fuel gauge, this variable resistor is connected to a float in the fuel tank. If the tank is filled with fuel, the sending unit resistance decreases, and the current flow through the bimetallic strip increases. This increased current flow heats the bimetallic strip and pushes the pointer toward the full position. A short to ground in the circuit between the sending unit and the gauge would also increase current and force the gauge to always read full, which is what answer C says.

Answer D is wrong. Anytime there is an open circuit the gauge would not work.

Question #54

Answer A is wrong. The voltage limiter produces a constant 5 V regardless of charging system voltage, an inoperative alternator does not affect the gauge reading.

Answer B is correct. Only Technician B is correct. An inoperative instrument voltage limiter may cause erratic gauge operation.

Answer C is wrong. Only Technician B is correct.

Answer D is wrong. Only Technician B is correct.

Question #55
Answer A is correct. Because the windows work normally through the master switch, we know the motors and their related circuits are fine. A is the best answer from the list of answer choices. If there is an open between the ignition switch and the window switch, the window switch cannot send voltage to the motor and the window will not work. Since the window switch circuit is separate from the master switch circuit, the window would operate through the master switch in spite of the problem.
Answer B is wrong. If there were an open in the window switch, the master switch would not move the window either.
Answer C is wrong. An open in the master switch ground wire would also make the window inoperative from the master switch.
Answer D is wrong. If there were a short to ground at the motor's circuit breaker, the window would not operate properly from either switch. The master control switch provides the overall control of the system. Power to the individual switches is provided through the master switch. The master switch may also have a safety lock switch to prevent operation of the windows by the individual switches. When the safety switch is activated, it opens the circuit to the other switches and control is only by the master switch. As an additional safety feature, some systems prevent operation of the individual switches unless the ignition switch is in the RUN or ACC position.

Question #56
Answer A is wrong. Only Technician B is correct.
Answer B is correct. Only Technician B is correct. When an electronic instrument cluster is completely inoperative, the instrument panel fuse should be tested before replacing the cluster.
Answer C is wrong. Only Technician B is correct.
Answer D is wrong. Only Technician B is correct.

Question #57
Answer A is wrong. A worn socket on the right rear lamp would not effect the left rear lamps operation.
Answer B is wrong. A faulty headlight switch would cause problems with both lamps.
Answer C is incorrect. Neither Technician is correct.
Answer D is correct. Neither Technician is correct.

Question #58
Answer A is wrong. Technician B is also correct.
Answer B is wrong. Technician A is also correct.
Answer C is correct. Both technicians are correct. The problem could be with the BCM. Occasionally the circuit which controls the clock will malfunction. Also, in some cases the clock will have its own separate fuse. In this case the fuse could be faulty.
Answer D is incorrect. Both technicians are correct.

Question #59
Answer A is correct. Only Technician A is correct. Regardless of the system, a digital speedometer depends on the input from a speed sensor to display the speed of the vehicle. If the sensor is faulty, as suggested by Technician A, the speedometer may not receive any speed inputs, and the gauge will read zero. Other results may be that the Malfunction Indicator Light (MIL) may be on, cruise control inoperative, and the ABS light on.
Answer B is wrong. The faulty TPS would result in poor engine running, but would not cause the speedometer to be inoperative.
Answer C is wrong. Only Technician A is correct.
Answer D is wrong. Only Technician A is correct.

Question #60
Answer A is wrong. A burned out bulb would cause this problem.
Answer B is correct. Warning lights are simple circuits. Power is supplied to the light at all times, and the sending unit switches to open or close the circuit by completing the path to ground. When satisfactory conditions are met, the circuit is open and the bulb is off. Any problem that opens the circuit will keep the light off at all times. Answers A, C, and D all suggest an open circuit and would cause the problem. Answer B will not cause the problem; if anything, it would cause the light to illuminate constantly. To test a faulty warning lamp on a system with a normally open switch (sending unit), turn the ignition switch to the START position. The prove-out circuit should light the warning lamp. If the light does not come on during the prove-out, disconnect the sender switch lead. Use a jumper wire to connect the sender switch lead to ground. With the ignition switch in the RUN position, the warning lamp should light. If the lamp is illuminated, test the prove-out circuit for an open. If the light does not come on, either the bulb is burned out or the wiring is damaged. Use a test light to confirm voltage is present at the sensor terminal connector. If there is voltage, the bulb is probably bad. If voltage is not present to the sending unit, the bulb may be burned out. At this point, the instrument cluster will need to be removed. With the cluster removed, check for battery voltage to the panel connector. If voltage is present, substitute a known good bulb and test again. If the system uses a normally closed switch, test in the same manner. However, there will not be a separate prove-out circuit.
Answer C is wrong. An open would cause the warning lights not to work.
Answer D is wrong. A defective sending unit would cause this problem.

Question #61
Answer A is wrong. Technician B is also correct.
Answer B is wrong. Technician A is also correct.
Answer C is correct. Both Technicians are correct. A burned out bulb and a faulty center high mounted stop light both could cause the driver information system to signal the operator that there is a stop light malfunctioning.
Answer D is wrong. Both technicians are correct.

Question #62
Answer A is correct. Only Technician A is correct. In the horn circuit given for the question, the horn is activated by depressing the horn switch that completes the relay circuit to ground. The relay is then activated and the horn sounds. If the horn sounds continuously, something is providing a ground for the relay or the relay contacts are stuck closed. Technician A correctly diagnosed the problem by saying the wire from the relay to switch must be grounded. This problem would bypass the horn button.
Answer B is wrong. If the relay points were stuck shut, the horn would sound continuously. If the relay points were stuck in an open position, the horn would never work. Technician B is wrong.
Answer C is wrong. Only Technician A is correct.
Answer D is wrong. Only Technician A is correct.

Question #63
Answer A is wrong. Neither technician is correct.
Answer B is wrong. Neither technician is correct.
Answer C is wrong. Neither technician is correct.
Answer D is correct. Neither technician is correct. If the rear defogger switch was faulty the rear defogger and the heated mirrors would not work. If the driver's side mirror had a faulty ground only the driver's side mirror would fail to work.

Question #64
Answer A is wrong. When servicing the air bag system, always disconnect the battery negative terminal first and wait for the manufacturer's specified time period to elapse. This time period is usually one or two minutes.
Answer B is wrong. Since deployed air bags may contain residual chemicals, wear safety glasses and gloves when handling these components. Always store inflator modules face-up on the bench, and carry these components with the trim cover facing away from your body.
Answer C is correct. Never use a powered test light to diagnose an air bag system. Diagnose these systems with a voltmeter or the manufacturer's recommended diagnostic tool(s). Use of an ohmmeter should be restricted to circuits without connections to pyrotechnic devices. Never use a test light to check the system.
Answer D is wrong. Handle all sensors with care. Do not strike or jar a sensor in such a manner that deployment may occur. Sensor brackets cannot be altered. They must maintain their original OEM configuration.

Question #65
Answer A is wrong. Technician B is also correct.
Answer B is wrong. Technician A is also correct.
Answer C is correct. Both Technicians are correct. Erratic cruise control operation may be caused by a leak in the vacuum hose connected to the servo or in the servo diaphragm.
Answer D is wrong. Both Technicians are correct.

Question #66
Answer A is wrong. A incorrectly suggests that the wiper motor would cause the problem. If the motor were bad, the wipers would not work properly at all times and in other modes of operation.
Answer B is correct. Only Technician B is correct. Speed sensitive wipers rely on input from the VSS to regulate speed. If there is no speed input, there is nothing to set the wiper speed to. Speed-dependent wipers compensate for extra moisture that normally accumulates on the windshield at higher speeds in the rain. At higher speeds, the delay between wipers shortens when the wipers are operating in the interval mode. This delay is automatically adjusted at speeds between 10 and 65 miles per hour. Basically, this system functions according to the input the computer receives about vehicle speed.
Answer C is wrong. Only Technician B is correct.
Answer D is wrong. Only Technician B is correct.

Question #67
Answer A is wrong. When testing an antenna with an ohmmeter, continuity should be present between the end of the antenna mast and the center pin on the lead-in wire.
Answer B is wrong. Continuity also should be present between the ground shell on the lead-in wire and the antenna mounting hardware.
Answer C is wrong. There should not be continuity between the center pin and the ground shell.
Answer D is correct. All of the statements are true, except D, which states there should be continuity between the end of the antenna mast and the antenna mounting hardware. If this were to happen, the entire vehicle would try to become the antenna and the result would be poor reception and static. Before removing the radio/component, do these simple checks to quickly determine whether the system problems are external. 1) Test the vehicle's radio system outside, not inside a building. Make sure the hood is down. 2) Most noise can be located on weak AM stations at the low frequency end of the tuning band. 3) Ignition noise on FM usually indicates a problem in the ignition system. 4) If a test antenna is going to be used, the base must be grounded to the vehicle's body. DON'T HOLD THE MAST. 5) Ninety percent of radio noise enters by way of the antenna. 6) Most "rubber" hoses (vacuum, coolant, etc.) are electrically conductive unless they have a white stripe. 7) When shielding hoses, wires, the dash, etc., use foil or screening material. Be sure to ground the material. 8) A weak or fading AM signal is normally caused by an improperly adjusted antenna trimmer (when used).

Question #68
Answer A is wrong. There is no need to check the multi-function switch, the pump is inoperative even when direct wired.
Answer B is wrong. There is no need to check the wiring harness, the pump is inoperative even when direct wired.
Answer C is correct. The pump is faulty and needs to be replaced. It will not operate when a known good power supply is attached to it.
Answer D is wrong. The pump is operative, there is no need to check the hoses and nozzles.

Question #69
Answer A is wrong. Technician B is also correct.
Answer B is wrong. Technician A is also correct.
Answer C is correct. Both technicians are correct. If the system does not disarm on the driver's side, move to the other side. If the system now disarms, the problem is in the driver's lock mechanism or circuit. One of the most difficult things about diagnosing an anti-theft system is that the circuits and wires are hidden. If they were not, someone could easily get around the system and defeat the purpose for having the system. The only way to identify the circuit and its components is by studying the wiring diagram and by using a component locator guide therefore, Technician B is also correct. Most antitheft systems offer self-diagnostic capabilities. Follow the service manual procedures for the proper method of entering diagnostics for the vehicle you are working on.
Answer D is wrong. Both Technicians are correct.

Question #70
Answer A is wrong. Answer A is not right. The indicator lamp would not illuminate if there were an open defogger relay winding. This would cause the defogger not to work, so it is half correct.
Answer B is wrong. An open circuit here would cause the indicator lamp and the heater grid not to work.
Answer C is correct. To identify the cause of this problem, look at the wiring diagram and identify the part of the circuit that controls the indicator light. The problem is not in that part of the circuit. The defogger part of the circuit contains the problem. Answer C describes a condition that would prevent the defogger from working but would allow the indicator to light.
Answer D is wrong. Answer D would prevent both the indicator and defogger from working. The control switch may be a three-position, spring-loaded switch that returns to the center position after making momentary contact to the ON or OFF terminals. Activation of the switch energizes the electronic timing circuit, which energizes the relay coil. With the relay contacts closed, direct battery voltage is sent to the heater grid. At the same time, voltage is applied to the ON indicator. The timer is activated for 10 minutes. At the completion of the timed cycle, the relay is de-energized and the circuit to the grid and indicator light is broken. If the switch is activated again, the timer will energize the relay for 5 minutes.

Question #71
Answer A is correct. Only Technician A is correct. If the alternator belt is bottomed in the pulley, the belt may slip and alternator output is reduced.
Answer B is wrong. A misaligned belt may cause rapid belt wear, but the belt should not slip.
Answer C is wrong. Only Technician A is correct.
Answer D is wrong. Only Technician A is correct.

Question #72
Answer A is correct. The cause of all of the locks not working in the lock mode must be somewhere in the circuit that is common to all of the doors. The cause could be the inability to disable the unlock mode or something that prevents the lock mode from being activated. Answer A is correct because an open ground at the unlock relay contacts would cause this problem. The lock motors get their ground through that connection.
Answer B is wrong. Answer B would affect only one lock and therefore is not a possible cause for the problem in the question.
Answer C is wrong. Answer C suggests an open in the circuit between the master switch and the lock relay coil. This may prevent master switch operation but not the door switch operation.
Answer D is wrong. Answer D would affect only the individual door lock switches and not the master. Some systems do not use relays. The switch provides control of current flow in the same manner as power seats or windows.

Question #73
Answer A is wrong. An oscilloscope converts electrical signals into a visual image representing voltage changes over a specific period of time. An upward movement of the trace does mean that the voltage has increased, and downward movements show a decrease in voltage. If the trace stays flat, the voltage is staying at that level.
Answer B is correct. Although the cleanness of the connection will ensure good and clean waveforms, the size and clarity of the trace is dependent on the voltage scale and the time reference selected on the scope. Selecting the incorrect scale can result in the trace being too small to see or too large to fit on the screen.
Answer C is wrong. A flat waveform does indicate a constant voltage level, such as a clean DC voltage signal.
Answer D is wrong. Because the trace is showing voltage over time, the movement of the trace represents time across the screen of the oscilloscope.

Question #74
Answer A is correct. Only Technician A is correct. If the chime works when the passenger door is opened but not when the driver's door is open the problem is likely the driver's door switch.
Answer B is wrong. If the chime module was faulty it MOST likely would not work for either door.
Answer C is wrong. Only Technician A is correct.
Answer D is wrong. Only Technician A is correct.

Question #75
Answer A is wrong. Answer A is incorrect because each mirror contains separate motors.
Answer B is correct. In the diagram, you will notice that voltage is supplied through a fuse to the power mirror switch assembly. When the mirror select switch is in the left position, it supplies voltage to the left mirror motor. When the left/right switch is pressed to the left position, a ground connection is completed from the left/right motor through the switch to the ground. Under this condition, the motor moves the left side mirror to the left. Similar action happens for all other directions and the right side mirror.
Answer C is wrong. The ground for the motors is at the switch.
Answer D is wrong. This is a 12-volt system. The motors need 12 volts to operate correctly.

Question #76
Answer A is wrong. Technician B is also correct.
Answer B is wrong. Technician A is also correct.
Answer C is correct. Both Technicians are correct. Static may be caused by the charging system or the ignition system. A poor engine ground or poor ground at the sound system components may cause static in the sound. Static can also result from poor grounds at the hood and other body components since they tend to help shield the Radio Frequency Interference (RFI) of the engine from the radio. Defective radio suppression devices, such as a suppression coil on an instrument voltage limiter or a clamping diode on an electromagnetic clutch, may cause static on the radio. A defective antenna with poor ground shielding may also result in static.
Answer D is wrong. Both Technicians are correct.

Question #77
Answer A is wrong. Technician A is wrong because a jammed track would open the circuit breaker only when the motor was energized and the seat was trying to move.
Answer B is correct. Only Technician B is correct. A grounded wire prior to the seat switches could cause the circuit breaker to cycle continuously. If the circuit breaker cycles without touching the seat control buttons, the fault that is causing the breaker to open must be before the seat control buttons.
Answer C is incorrect. Only Technician B is correct.
Answer D is incorrect. Only Technician B is correct.

Question #78

Answer A is wrong. A defective stator could cause a whine noise.

Answer B is wrong. A defective diode could cause voltage spikes, thus creating a whine noise.

Answer C is wrong. A defective capacitor could cause a whine noise.

Answer D is correct. All of the answer choices in this question are true, with the exception of D, an open field winding. This problem would tend to decrease the chance of radio noise because the alternator (a major source of noise) would not work. If the stator, a diode in the alternator, or a noise suppression capacitor were defective, these could be the source of the noise. A defective alternator will usually cause a whining noise. A whining noise can be caused by shorted diodes or stator windings, or by a dry rotor bearing. A quick way to test for the cause of a whining sound is to disconnect the wiring to the generator, then start and run the engine. If the noise is not there, the cause of the noise is a magnetic whine due to shorted diodes or stator windings. Use a scope to verify the condition of the diodes and stator. If the noise remains, the cause is mechanical and probably due to worn bearings.

Question #79

Answer A is wrong Technician B is also correct.

Answer B is wrong. Technician A is also correct.

Answer C is correct. Both Technicians are correct. Slow moving or dimly operating electrical parts are typically caused by excessive resistance. The increased circuit resistance decreases the current through the circuit causing the motor to run slowly or a bulb to burn dimmer. This is the logic used by Technician B, who is correct. However, when we have a motor that is performing some mechanical function, there is a possibility that mechanical things are preventing the motor from turning at its normal speed. This is what Technician A suggests and he is also correct. To determine which of these is the cause of the problem, you can measure the current flow in the circuit. If the problem is excessive electrical resistance, the circuit's current will be low. If the problem is mechanical, the current will be higher than normal due to the slow rotating motor.

Answer D is wrong. Both Technicians are correct.

Question #80

Answer A is wrong. This high voltage would cause an overcharged battery.

Answer B is wrong. The high voltage could cause damage to electronic components.

Answer C is wrong. With an overcharged battery, gassing does increase and answer C is also right. Overcharge conditions can usually be traced to voltage regulator circuit failures.

Answer D is correct. The high system voltage would increase the brilliance of the lights, not decrease it. Obviously 16.2 volts from the charging system is too much for a 12-volt battery.

Answers to the Test Questions for the Additional Test Questions Section 6

1.	C	12.	C	23.	D	34.	C
2.	B	13.	A	24.	C	35.	C
3.	B	14.	A	25.	C	36.	D
4.	C	15.	B	26.	A	37.	C
5.	C	16.	C	27.	A	38.	A
6.	B	17.	B	28.	A	39.	D
7.	B	18.	C	29.	C	40.	B
8.	B	19.	B	30.	B	41.	D
9.	A	20.	A	31.	B	42.	C
10.	D	21.	C	32.	C	43.	B
11.	C	22.	A	33.	A	44.	B

Explanations to the Answers for the Additional Test Questions Section 6

Question #1

Answer A is wrong. Technician B is also correct.

Answer B is wrong. Technician A is also correct.

Answer C is correct. Both Technicians are correct. Technician A is right, but answer A is wrong since both technicians are right. If a cruise-control circuit loses the input from the VSS, the system will not work. Logic explains why. If the control module does not sense a speed, how can it set a speed? Technician B is also right, but answer B is wrong. If the gauge fuse is bad, the cruise-control module will be affected. It will not have any power to it. If the voltage signal from the VSS drops below the low comparator value, the control module energizes the supply valve solenoid to allow more vacuum into the servo and increases the throttle opening. When the VSS signal returns to a value within the comparator levels, the supply valve solenoid is de-energized. If the VSS signal is greater than the high comparator value, the control module de-energizes the vent solenoid valve to release vacuum in the servo. The vehicle speed is reduced until the VSS signals are between the comparator values, at which time the control module will energize the vent valve solenoid again. This constant modulation of the supply and vent valves maintains vehicle speed. Some vehicles now use electric motors to operate the throttle instead of vacuum.

Answer D is wrong. Both Technicians are correct.

Question #2

Answer A is wrong. Excessive internal resistance in the motor would cause the current to decrease.

Answer B is correct. Only Technician B is correct. The circuit breaker that protects an electric window circuit blows whenever an attempt is made to lower the window. Whenever a circuit breaker blows, think of high current. Normally an increase in current is caused by a short. But higher than normal current draw by a motor can be caused by the motor's inability to rotate at its normal speed. B is correct. If the window regulator is sticking, the motor would not be able to rotate at normal speed and current would be higher. Testing and repair of the power window system will usually require the door panels to be removed. There are several methods used by manufacturers to secure the door panel. Always refer to the proper service information to determine the correct methods of removal and installation of the panel to prevent damage. Also use new clips to assure a tight connection and eliminate noise from the panel. Most doors will have a sound-dampening material behind the panel. Therefore, you must remember to reinstall this material.

Answer C is wrong. Only Technician B is correct.

Answer D is wrong. Only Technician B is correct.

Question #3

Answer A is wrong. Total resistance is found by adding the resistors.

Answer B is correct. This is a simple Ohm's law problem. Since all of the resistors are in series, the answer is 12 Ohms. The other answers are incorrect because they do not add the resistors.

Answer C is wrong. Total resistance is found by adding the resistors.

Answer D is wrong. Total resistance is found by adding the resistors.

Question #4
Answer A is wrong. Technician B is also correct.
Answer B is wrong. Technician A is also correct.
Answer C is correct. Both Technicians are correct. When the alternator voltage is erratic or too low, the alternator may be full-fielded to determine the cause of the problem. When the alternator is full-fielded and the alternator current and voltage output are normal, the voltage regulator is probably defective. If the charging system voltage is higher than specified, the voltage regulator is probably defective. When the output is zero during the test, the field circuit is probably open. Worn brushes or an open field winding in the rotor usually causes this problem. If output is less than specified, there is probably a problem with the diodes or stator. A high resistance in the field winding also reduces output. When the alternator is full-fielded to test output, the voltage regulator is bypassed and does not affect output. Never full-field an alternator longer than 30 seconds. When full-fielding the system, the battery should be loaded to protect vehicle electronics and computers. With the voltage regulator bypassed, there is no control of voltage output. The AC generator is capable of producing well over 30 volts. This increased voltage will damage the circuits not designed to handle that high of voltage.
Answer D is wrong. Both Technicians are correct.

Question #5
Answer A is wrong. Both A and B are correct.
Answer B is wrong. Both A and B are correct.
Answer C is correct. Both A and B are approved methods of disarming the air bag system. The only disadvantage of disconnecting the battery is that the memory of accessory items like the entertainment system will be lost.
Answer D is wrong. Both A and B are correct.

Question #6
Answer A is wrong. A voltmeter connected across the input and output terminals of an instrument cluster illumination lamp rheostat displays 12.6 volts with the switch in the maximum brightness position and the engine off. This means the voltage drop across the rheostat is 12.6 volts and the voltage available at the lamps will be 0.0 volts. Answer A would be correct if the control were working properly, which it apparently is not.
Answer B is correct. With a voltage drop of battery voltage across the rheostat, no voltage would be available to the lamps.
Answer C is wrong. Answer C states that the rheostat is working normally; this is wrong.
Answer D is wrong. There is enough information to solve the problem. The power source for the instrument panel lights is provided through the headlight switch. The contacts are closed when the headlight switch is located in the PARK or HEADLIGHT position. The current must flow through a variable resistor (rheostat) that is either a part of the headlight switch or a separate dial on the dash. The resistance of the rheostat is varied by turning the knob. By varying the resistance, changes in the current flow to the lamps control the brightness of the lights.

Question #7
Answer A is wrong. A faulty hazard switch could cause a problem with brake lights, however, there would MOST likely be a problem with the hazard flasher operation as well, and the stem of the question states everything else works normally.
Answer B is correct. Only Technician B is right. A faulty brake switch could cause this problem. The brake lights are operated by the brake switch.
Answer C is wrong. Only Technician A is correct.
Answer D is wrong. Only Technician A is correct.

Question #8
Answer A is wrong. If the ground was bad on the control side of the horn relay the horn would either work or not work, because the relay would either energize or not energize.
Answer B is correct. If the horn ground was faulty the horn may sound, however it may weak due to lower than normal current flow.
Answer C is wrong. A faulty power supply to the horn switch would cause the horn to fail to work.
Answer D is wrong. A poor ground on the horn switch would cause the horn not to work.

Question #9
Answer A is correct. Only Technician A is correct. An open in circuit 58 could cause the low speed to be inoperative.
Answer B is wrong. An open in circuit 63 would turn off the whole system including the washer pump. When the switch is placed in the LOW SPEED position, battery voltage is supplied through wiper 1 to the low-speed brush. Wiper 2 also moves to the LOW position, but does not complete any circuits. Current flows through the armature, the low-speed brush, and the common brush to ground. Park switch position has no effect on motor operation.
Answer C is wrong. Only Technician A is correct.
Answer D is wrong. Only Technician A is correct.

Question #10
Answer A is wrong. If the filaments were shorted, there would be 12 volts available at the bulbs. Also, it is very unlikely that both bulbs would have the same problem.
Answer B is wrong. If there were an open between one bulb and ground, only that one bulb would be affected.
Answer C is wrong. There should be low resistance in the switch. It is a conductor.
Answer D is correct. The problem in question is simply that there is no voltage being applied to the backup lamps. The only possible cause listed in the answers is an open wire between the switch and the lamps. High resistance in a bulb circuit would cause dim operation. Considering all of these facts, answers A, B, and C are wrong.

Question #11
Answer A is wrong. Technician B is also correct.
Answer B is wrong. Technician A is also correct.
Answer C is correct. Both Technicians are correct. The meter shown for this question is measuring the voltage drop across the control circuit of a motor. Ideally there would be zero volts dropped. Because there is a large amount of voltage dropped across this part of the circuit, the motor would turn very slowly. Both identified a possible source for this unwanted resistance.
Answer D is wrong. Both Technicians are correct.

Question #12
Answer A is wrong. 9V would mean there is too much voltage drop in the circuit.
Answer B is wrong. 9.5V would mean there is to much voltage drop in the system.
Answer C is correct because 12 volts should be available before the horn. All the other answers show voltage drops greater than 1.2 volts; therefore, they are incorrect. All wiring must have resistance values low enough to allow enough voltage to the load for proper operation. The maximum allowable voltage loss due to voltage drops across wires, connectors, and other conductors in an automotive circuit is 10 percent of the system voltage. Therefore, in a 12-volt automotive electrical system, this maximum total loss is 1.2 volts.
Answer D is wrong. 6V would mean there is too much voltage drop in the system.

Question #13
Answer A is correct. Only Technician A is correct. The major cause of a fuse blowing is a short; therefore, it is highly likely that the cause of the problem is a short in the fuse's circuit. Fuse 17 is the true fuse for this sunroof circuit. If the power saver relay winding is shorted to ground, the fuse will blow. Therefore Technician A is correct.
Answer B is wrong. If the short is between the power saver relay winding and ground, all we have is a redundant ground and the fuse will not blow. Technician B is incorrect. Fuses do not wear out. They fail because something went wrong. Never replace a fuse or fusible link, or reset a circuit breaker, without finding out why it opened.
Answer C is wrong. Only Technician A is correct.
Answer D is wrong. Only Technician A is correct.

Question #14

Answer A is correct. A good diode will always have low resistance when an ohmmeter is connected across it in one direction and high resistance when the leads are reversed.

Answer B is wrong. If the meter indicates two infinite readings, the diode is open and damaged.

Answer C is wrong. If the meter readings are low in both directions, the diode is shorted.

Answer D is wrong. The numbers given in answer D are not good examples of actual readings. The 40 ohms is not considered high when checking a diode. The high reading would more likely be 40K ohms. You may run into problems when checking a diode with a high-impedance digital ohmmeter; use an analog meter if available. Since many diodes won't allow current flow through them unless the voltage is at least 0.6 volts, a digital meter may not be able to forward bias the diode. This will result in readings that indicate the diode is open, when in fact it may not be. Because of this problem, many multimeters are equipped with a diode testing feature. This feature allows for increased voltage at the test leads. The value displayed is the voltage required to forward bias the diode. If the diode is open, the meter will display "OL" or another reading to indicate infinity or out of range. Some meters during diode check will make a beeping noise when there is continuity.

Question #15

Answer A is wrong. Technician A suggests that low readings could result from a short to ground in the wire between the sending unit and the gauge. If there were a short here, the gauge would read high because the hot coil would have a very strong magnetic field and would pull the pointer toward it.

Answer B is correct. Only Technician B is correct. Some gauges contain two coils, and the pointer is mounted on a magnet under these coils. In a temperature gauge, the sending unit is connected to the hot coil and the cold coil is grounded. If the coolant is cold, the sending unit has a high resistance. Under this condition, current flows through the lower resistance of the cold coil. Coil magnetism around the cold coil attracts the magnet and the pointer to the cold position. As the coolant temperature increases, the sending unit resistance decreases. When the engine is at normal operating temperature, the current flows through the lower resistance of the hot coil and sending unit. This action attracts the magnet and the pointer to the hot position. When the gauge reads low, the problem could be a low magnetic field around the hot coil. Technician B says the wire from the gauge to ground may be open. If this were the case, the gauge would never move away from cold because the hot coil would never be energized.

Answer C is wrong. Only Technician B is correct.

Answer D is wrong. Only Technician B is correct.

Question #16

Answer A is wrong. High resistance causes low current draw.

Answer B is wrong. High resistance causes low current draw.

Answer C is correct. Again, sometimes there are easy questions. To answer this, think about what could cause a fuse to blow. The obvious answer is high current flow, which results from a short. Of the answer choices only C suggests a short and therefore can be the only cause of the problem.

Answer D is wrong. Answer D comes close, but this is a short on the ground side. It will not affect current flow. The short just becomes a redundant ground. The other answers focus on high resistance. High resistance always decreases current flow, and low current will never cause a fuse to blow. The technician should look for possible contamination or a foreign object lodged in the lighter.

Question #17

Answer A is wrong. Low resistance causes an increase in current flow.

Answer B is correct. Anytime there is a decrease in current flow, there is an increase in resistance.

Answer C is wrong. With a short to ground comes high current, so answer C cannot be right.

Answer D is wrong. And answer D is wrong because there is no current flow when the circuit is open. With the circuit drawing less amps than normal, the bulb would be dim.

Question #18

Answer A is wrong. Technician B is also correct. Answer A must be wrong since the motor and circuit works to take the top down.

Answer B is wrong. Technician A is also correct. Answer B is wrong because an open ground wire at the switch would stop the top from going up or down.

Answer C is correct. Both Technicians are correct. This system uses electric circuits to move mechanical linkages. Therefore both need to be considered as the cause of the problem and both need to be checked. On the electrical side of things, diagnosis should be focused on the "up" side of the circuit. Answer C relates to the linkage. Since moving the top requires more strain on the system, it is very likely that a binding linkage will be more noticeable when the top is moving up rather than down. So answer C is a likely cause of the problem.

Answer D is wrong. Both Technicians are correct. Answer D suggests a problem that does not allow the top to move up or down. Convertible top linkage needs to be lubricated at OEM-specified intervals.

Question #19

Answer A is wrong. A wire shorted to ground between the turn signal light switch and the right front signal light would cause the right turn signals to operate abnormally.

Answer B is correct. Only Technician B is correct. High resistance between the LF signal and ground could cause the LH turn signals to flash rapidly.

Answer C is wrong. Only Technician B is correct.

Answer D is wrong. Only Technician B is correct.

Question #20

Answer A is correct. The clue to answering this question is to think about what is happening. A normal light bulb in a normal circuit will only burn brighter if more voltage is applied to it. This is exactly what A says. Higher charging system voltage will cause the bulb to be brighter.

Answer B is wrong. High resistance in the circuit would cause the lights to be dimmer.

Answer C is wrong. A shorted diode would decrease the output of the charging system and would probably result in a somewhat dimmer bulb.

Answer D is wrong. The same is true for high resistance in the battery wire from an alternator, so answer D is also wrong. If the charging voltage was too high, there may be a problem in the following areas: 1) Defective voltage regulator. 2) Poor voltage regulator ground connection. 3) Short to ground in the field coil control circuit causing the AC generator to full field. 4) High resistance in the "sense" circuit between the battery and the PCM or voltage regulator.

Question #21

Answer A is wrong. Answer A suggests a faulty master switch. The switch is good because there are 12 volts at Terminal 3.

Answer B is wrong. Answer B states the window switch may be faulty. This also could not be right because there are 12 volts at Terminal 5.

Answer C is correct. The power window motor shown in the figure is completely inoperative. Different voltage measurements are given for terminals of the master window switch placed in the "down" position. To answer this question, go through each answer and check the validity by looking at the voltages given. The switch is completing the path for the down mode. Answer C suggests that the motor is faulty. We know this is possible because there are 12 volts applied to the motor and the ground circuit is complete. It has everything it needs to work. It just is not working.

Answer D is wrong. Answer D suggests a poor circuit ground. If this were the case, the motor would work in all modes but would be sluggish.

Question #22

Answer A is correct. Only Technician A is correct. Circuit 752 controls the ground-side of the medium/high blower circuit. If there were an open in that circuit it would only affect the medium/high speed.

Answer B is wrong. The middle resistor in the block is for the low and medium/low speeds only. If the middle resistor was open, low, medium-low, and medium-high would all be inoperative.

Answer C is wrong. Only Technician A is correct.

Answer D is wrong. Only Technician A is correct.

Question #23
Answer A is wrong. Neither Technician is correct.
Answer B is wrong. Neither Technician is correct.
Answer C is wrong. Neither Technician is correct.
Answer D is correct. Neither Technician is correct. The instrument voltage regulator (IVR) supplies about 5V to the gauges regardless of the charging system voltage. Technician A is wrong; a good IVR will not be affected by the charging system. Technician B is also wrong. Sending units should be tested with an ohmmeter. Bypassing them with a jumper wire may damage the gauge, whether or not there is an IVR in the circuit. Most sending units range from high to low resistance. Jumping across it would put zero resistance into the circuit and the gauge would receive high current. The proper way to perform this type of test is to use a variable resistor between the disconnected sending unit wire and ground. As the resistance is changed, the needle should move on the gauge.

Question #24
Answer A is wrong. Technician B is also correct.
Answer B is wrong. Technician A is also correct.
Answer C is correct. Both Technicians are correct. Looking at the Rd/Lt Gn wire near the motor, you will notice it is the power feed to the motor when the switch activates the "up" function. There should be 12 volts there, which is what A said. Since that function of the system does not work, there may not be 12 volts there. You will also notice that the Yl/Lt Gn wire connects and completes the circuit from the circuit breaker at the motor to the up switch. If this circuit were open, the seat would not move up or down. The power seat system is classified by the number of ways in which the seat is moved. The most common classifications are: 1) Two-way: moves the seat forward and backward. 2) Four-way: moves the seat forward, backward, up, and down. 3) Six-way: moves the seat forward, backward, up, down, front tilt, and rear tilt.
Answer D is wrong. Both Technicians are correct.

Question #25
Answer A is wrong. An open fuse would prevent all radio operation.
Answer B is wrong. A blown antenna fuse would not allow the antenna to go up.
Answer C is correct. To answer this question, think about what is happening in this power antenna circuit. When the radio is turned on, voltage is supplied to the relay winding. This action moves the relay points to the up position, and current flows through the motor to move the antenna upward. When the antenna is fully extended, the up limit switch opens and stops the current flow through the motor. When the radio is turned off, current flow through the relay coil stops. Under this condition, the relay contacts move to the down position. This action reverses current flow through the motor and moves the antenna downward. When the antenna is fully retracted, the down limit switch opens and stops the current flow through the motor. Of the answer choices, only the down limit switch could cause the problem. If the down limit switch is open, the motor will never be activated to bring the antenna down. The other answers would affect the total operation of the antenna.
Answer D is wrong. A shorted antenna motor would blow the fuse and the motor would not work in either direction.

Question #26
Answer A is correct. Only Technician A is correct. Referring to the diagram for this question, you will see that A is correct. An open winding in the horn relay would prevent the horn from working. If the winding were open, the relay would not close its contacts to send voltage to the horns.
Answer B is wrong. Technician B is wrong. A shorted horn switch would cause the horns to sound continuously. When a customer complains of no horn operation, first confirm the complaint by depressing the horn button. If it is mounted in the steering wheel, rotate the steering wheel from stop to stop while depressing the horn button. If the horn sounds intermittently while the steering wheel is turned, the problem is probably in the sliding contact ring in the steering column, or the tension spring is worn or broken. If the horn does not sound during this test, continue to check.
Answer C is wrong. Only Technician A is correct.
Answer D is wrong. Only Technician A is correct.

Question #27

Answer A is correct. This question contains a summary of all of the bulb circuits mentioned in the question. Typically the turn signal and brake lights share a common bulb filament.

Answer B is wrong. Answer B is wrong because backup lamps normally have a single filament bulb in its own housing.

Answer C is wrong. Answer C suggests that the taillights and stoplight share a filament. This is not true; the brake lights operate brighter than the taillights and use a separate filament.

Answer D is wrong. Answer D is wrong because the brake pedal switch closes the circuit when the pedal is depressed.

Question #28

Answer A is correct. Anytime a fuse blows, suspect a short in the affected circuit. Answer A refers to a short, but it is in the stoplight circuit, so it is unlikely that this would cause the taillight fuse to blow.

Answer B is wrong. Answer B is an open and opens do not blow fuses.

Answer C is wrong. Answer C is an intermittent short to ground in the taillight wire; this would definitely blow the fuse whenever the wire contacts ground.

Answer D is wrong. Answer D would cause intermittent failure of the lights but would not cause the fuse to blow because it is not a short, it would be an open. The most commonly used circuit protection device is the fuse. A fuse contains a metal strip that will melt when the current flowing through it exceeds its rating. The thickness of the metal strip determines the rating of the fuse. When the metal strip melts, excessive current is indicated. The cause of the overload must be found and repaired; then a new fuse of the same rating should be installed. The most commonly used automotive fuses are rated from 4 to 30 amps.

Question #29

Answer A is wrong. Technician B is also correct.

Answer B is wrong. Technician A is also correct.

Answer C is correct. Both Technicians are correct. If the door ajar switches are corroded or damaged, they cannot complete the circuit telling the anti-theft controller that the doors are open. Likewise, most vehicles have circuits that turn on the interior lights when a door is opened. If the lights do not come on when the door is opened, you know the switch is bad.

Answer D is wrong. Both Technicians are correct.

Question #30

Answer A is wrong. Always disconnect the negative cable first. This prevents sparks if the wrench touches the frame or other body component while removing the first cableTherefore, Technician A is wrong and is thinking dangerously. Whenever you disconnect a battery, you should disconnect the negative cable first. Failure to do this can cause damage to the vehicle's components and you.

Answer B is correct. Only Technician B is correct. Maintenance-free batteries with built-in hydrometers indicate a low charge (below 65 percent full charge) when the hydrometer is dark. The sight glass will appear green if the battery is over 65 percent charged.

Answer C is wrong. Only Technician B is correct.

Answer D is wrong. Only Technician B is correct.

Question #31

Answer A is wrong. Answer A has two things wrong: the test is not measuring the positive cable, and the voltage drop is not excessive. Answer A is incorrect because resistance is not excessive, it is normal.

Answer B is correct. The figure shows the hook-up for conducting a voltage drop test across the negative battery cable, from the battery to the ground at the starter. The typical maximum allowable voltage drop across this cable is 0.1 volts.

Answer C is wrong in the same way; it is the ground cable being measured.

Answer D is wrong. Answer D has the right test and the wrong cable and has declared the voltage drop to be excessive when it is not.

Question #32

Answer A is wrong. Answer A says the voltage drop from the groundside of the grid to an engine ground should not exceed one volt. This is not true. Battery voltage should be dropped by the grid and zero volts dropped at the ground. If there is a voltage drop at the ground, we have a bad ground.

Answer B is wrong. Answer B says a 12-volt test light should be illuminated at half brilliance at any place on the grid. This is false. The brilliance of the test light should decrease as it moves from the positive side to the negative, simply because voltage is being dropped along the grid.

Answer C is correct. Answer C states the test light will not illuminate on part of a grid if there is an open. This is true. Another indication of an open grid (besides the fact that the window will not be cleared around that grid) is that the test light will have full brilliance before the open. After the open, the test light will not come on. Between the two is the location of the open.

Answer D is wrong. Answer D says the opposite of what to expect. The brilliance of the test light decreases as it moves toward the negative or ground side of the grid. When electrons are forced to flow through a resistance, heat is generated. Rear window electric defoggers use this principle of controlled resistance to heat the glass. The resistance is through a grid that is baked on the inside of the glass. The terminals are soldered to the vertical bus bars. One terminal supplies the current from the switch; the other provides the ground.

Question #33

Answer A is correct. Only Technician A is correct. The rheostat in the headlight switch determines the brightness of the instrument cluster bulbs. It is connected in series with the instrument cluster bulbs. Since it is in series, an open rheostat will open the circuit to the instrument cluster bulbs.

Answer B is wrong. The instrument cluster bulbs are connected in parallel. If one bulb burns out or opens, the other bulbs remain illuminated. The instrument cluster bulbs are connected in parallel. If one bulb burns out or opens, the other bulbs remain illuminated. The power source for the instrument panel lights is provided through the headlight switch. The contacts are closed when the headlight switch is located in the PARK or HEADLIGHT position. The current must flow through a variable resistor (rheostat) that is either a part of the headlight switch or a separate dial on the dash. The resistance of the rheostat is varied by turning the knob. By varying the resistance, changes in the current flow to the lamps control the brightness of the lights.

Answer C is wrong. Only Technician A is correct.

Answer D is wrong. Only Technician A is correct.

Question #34

Answer A is incorrect. Technician B is also correct.

Answer B is incorrect. Technician A is also correct.

Answer C is correct. Both Technicians are correct. If the contacts at the RR window switch are stuck or remain open, the motor will never be activated. We know parts of the circuit are good because the LR window works normally. Diagnosis should begin by determining what type of electrical problem we have. Since the window does not work, an open is the MOST likely problem. Both technicians have identified this as the basic problem. Now, if we look at the wiring diagram and note where the techs describe the open and then determine what the effect would be, we will be able to see which tech is right. They both are. An open between the RR window switch and ground would remove the ground from the motor and it will not be able to work. A test light or DMM would work equally well in diagnosing this type of failure.

Answer D is wrong. Both Technicians are correct.

Question #35

Answer A is wrong. Technician B is also correct.

Answer B is wrong. Technician A is also correct.

Answer C is correct. Both Technicians are correct. A weak or damaged solenoid plunger return spring may cause the starter to run with the engine for a few seconds, or longer when the engine starts. A misaligned starting motor may also cause this problem.

Answer D is wrong. Both technicians are correct.

Question #36
Answer A is wrong. A is wrong. If the circuit was repaired correctly, there should be zero volts after the motor. Battery voltage should be dropped by the rest of the circuit.
Answer B is wrong. If everything was repaired correctly, the test light should illuminate with full brilliance at the power feed to the motor. The feed circuit before the motor will drop very little voltage. If either condition existed as described by these technicians, the motor would not develop full speed.
Answer C is wrong. Neither Technician is correct.
Answer D is correct. Neither Technician is correct.

Question #37
Answer A is wrong. Technician B is also correct.
Answer B is wrong. Technician A is also correct.
Answer C is correct. Both Technicians are correct. When replacing a composite or halogen bulb, always turn off the lights and allow the bulbs to cool before proceeding. Also, keep moisture away from the bulb and handle the bulb only by its base. Do not scratch or drop the bulb. Coat the terminals of the bulb or at the connector with dielectric grease to minimize corrosion. Be careful not to get the grease on the bulb. By using the composite headlight system, vehicle manufacturers are able to produce any style of headlight lens they desire. This improves the aerodynamics, fuel economy, and styling of the vehicle. Many manufacturers vent the composite headlight housing because of the increased amount of heat developed by these bulbs. Because the housings are vented, condensation may develop inside the lens assembly. This condensation is not harmful to the bulb and does not affect headlight operation. When the headlights are turned on, the heat generated from the halogen bulbs will dissipate the condensation quickly. Some manufacturers use integrated nonvented composite headlights. On these vehicles, condensation is not considered normal. The assembly should be replaced.
Answer D is wrong. Both Technicians are correct.

Question #38
Answer A is correct. Only Technician A is correct. When a motor rotates at a slower speed than normal, it will draw more current than normal. Therefore, we have a basic explanation of what is happening to the starter in this question. Now the question is what is causing it. Technician A suggests that the starter has worn bushings. This is quite possible. If the armature cannot rotate freely, it will rotate slower and current will increase.
Answer B is wrong. Technician B states there may be high resistance in the battery ground cable. This would not cause a higher than normal current draw. High resistance in the cable would cause lower than normal current draw.
Answer C is wrong. Only Technician A is correct.
Answer D is wrong. Only Technician A is correct.

Question #39
Answer A is wrong. There would be no change in operation.
Answer B is wrong. The backup lights would work through the redundant ground.
Answer C is wrong. Because the ground is redundant, all of the other individual sets of lights will work normally. All the other answers are wrong. However, if the left ground becomes overloaded, the result will be a voltage drop on the left ground, which could cause the lights to operate dimly.
Answer D is correct. Following the wires in the diagram you will notice that the ground for the lights is redundant. Therefore, if one is bad, the other will complete the circuit. For this reason, answer D is correct.

Question #40
Answer A is wrong. Answer A suggests the problem is a shorted close switch. Since the problem is the inability of the roof to close, the short cannot cause the problem. The switch is designed to close the circuit by supplying a ground. If the ground were always there, the roof would not open.
Answer B is correct. To answer this question, you again need to locate the wires and components on the wiring diagram given with the question. Answer B may be correct. An open in the close relay control circuit may stop the roof from closing.
Answer C is wrong. Answer C would shut down the whole system.
Answer D is wrong. This would shut down the whole circuit.

Question #41
Answer A is wrong. If the starter motor is not working and there is 12-volt drop across the solenoid, an open between the solenoid and the starter, or within the starter, is the MOST likely cause. Technician A is wrong because there is no voltage drop.
Answer B is wrong. Technician B is also wrong because the starter solenoid is engaging.
Answer C is wrong. Neither Technician is correct.
Answer D is correct. Neither Technician is correct.

Question #42
Answer A is wrong. Technician B is also correct.
Answer B is wrong. Technician A is also correct.
Answer C is correct. Both Technicians are correct. In order for an alternator to have output, it must be rotating and it must have field current. If there is an open in field circuit, no field voltage will be available and the unit will have zero output. Likewise, if the fuse or fusible link in the alternator to battery wire is open, the alternator will have zero output. To properly test and service the charging system, it is important to identify the field circuit being used. Automobile manufacturers use three basic types of field circuits. The first type is called the A circuit. It has the regulator on the ground side of the field coil. The B+ for the field coil is picked up from inside the AC generator. By placing the regulator on the ground side of the field coil, the regulator will allow the control of field current by varying the current flow to ground. The second type of field circuit is called the B circuit. In this case, the voltage regulator controls the power side of the field circuit. Also, the field coil is grounded from inside the AC generator. The third type of field circuit is called the isolated field. The AC generator has two field wires attached to the outside of the case. The voltage regulator can be located on either the ground (A circuit) or on the B+ (B circuit) side. Regardless of which type is used, the field circuit is designed to control the amount of voltage output by controlling the amount of current through the field windings. As rotor speed increases, field current must be decreased to maintain regulated voltage.
Answer D is wrong. Both Technicians are correct.

Question #43
Answer A is wrong. Usually the current for the hazard circuit is not carried by the turn signal flasher.
Answer B is correct. The hazard lights use basically the same circuit as the turn signals. The difference is in the control. Turn signals flash just at the side of the vehicle, and the hazard system flashes at all four corners. If there is a problem with the flashers and not with the turn signals as well, the problem can be in that part of the circuit that is only part of hazard light control. Of the answers given in this question, only B is unique to the hazards. Therefore, only it can be the cause of the problem. The other answers would affect turn signal operation, which work fine. The hazard-warning system is part of the turn signal system. It has been included on all vehicles sold in North America since 1967. All four turn signal lamps flash when the hazard-warning switch is turned on. Depending on the manufacturer, a separate flasher can be used for the hazard lights than the one used for the turn signal lights. The operation of the hazard flasher is identical to that of the turn signal. The only difference is that the hazard flasher is capable of carrying the additional current drawn by all four turn signals. And, it receives its power source directly from the battery.
Answer C is wrong. An open circuit here would affect the turn signals.
Answer D is wrong. If the signal light switch was defective the turn signals would not work properly.

Question #44
Answer A is wrong. An engine that has low compression would crank fast but still draw low current. This is because the starter will rotate freer.
Answer B is correct. Only Technician B is correct. The vehicle in this question has a slow crank and low current draw problem. Typically high starter current draw and low cranking speed usually indicates a defective starter. This condition may also be caused by internal engine problems, such as partially seized bearings. Low current draw and slow cranking speed usually indicates excessive resistance in the starter circuit, just as B suggests.
Answer C is wrong Only Technician B is correct.
Answer D is wrong. Only Technician B is correct.

Glossary

Air bag A passive restraint system having an inflatable bag located in the center of the steering wheel in front of the driver and, in later model vehicles, a second inflatable bag located in the dash in front of the front seat passenger that inflates on vehicle impact.

Alternator A belt-driven generator that converts mechanical energy to electrical energy.

Antenna A wire or other conductive device used for radiating or receiving electromagnetic signals, such as those for radio, television, or radar.

Anti-theft system A deterrent system designed to scare off would be vehicle thieves.

Armature A part moved through a magnetic field to produce an electric current.

Backup light Lamps that provide rear illumination when the vehicle is being backed up.

Battery A device that converts chemical energy into electrical energy.

Bearing A component that reduces friction between a stationary and rotating part, such as a shaft.

Body control module A component of the computerized self diagnosis system.

Brake lights Red lamps at the rear of the vehicle that are illuminated when the brake pedal is applied.

Brake-light switch A component of the braking system that completes an electrical circuit to illuminate the brake lights when the brake pedal is applied.

Brush An conductive component that rides on the commutator or slip ring to provide an electrical circuit between rotating and stationary components.

Bulb A glass envelope containing a filament to provide illumination.

Bulkhead connector A connector for wires that are to pass through the bulkhead of a vehicle.

Buzzer An electric sound generator that makes a buzzing noise.

Capacitor An electrical device for the temporary storage of electricity, often used to reduce RFI.

Charge The passing of an electric current through a battery to restore its energy.

Concealed headlights A headlamp system that retracts the lamps into the bodywork when they are turned off.

Corrosion A chemical action that eats away material such as metal, paint, or wire.

Courtesy light Lamps that illuminate the interior of a vehicle when a door is opened.

Cruise control A device that automatically maintains vehicle speed over a wide range of terrain conditions.

Current The flow of electricity, measured in amperes.

Current-draw test A test to determine amperes required by the starter motor during starting operation.

Defogger A part of a heater system to prevent windshield or rear window fogging or icing.

Digital An electrical signal having two states, on and off.

Electric fuel pump An electrical device used to draw fuel from the fuel tank and deliver it to the engine.

Electronic-ignition system An ignition system controlled by solid state electrical signals.

Field coil A coil of insulated wire, usually wound around an iron core, through which current is passed to produce a magnetic field.

Flywheel A heavy metal wheel that is attached to the crankshaft and rotates with it.

Fusible link A bar or wire that is designed to melt due to heat of a specific current passing through it is exceeded.

Ground The path, generally the body of the vehicle, for the return of an electrical circuit. Also, a term used for causing an accidental or intentional short circuit.

Harness A group of electrical conductors.

Hazard-warning system Vehicle perimeter lighting and associated switches and wiring that flash giving warning to a hazard.

Headlamps The lamps at the front of a vehicle to provide illumination for the road ahead.

Horn relay An electromagnetic device used to activate the horn when the horn switch is closed.

Hydrometer An instrument used to measure the specific gravity of a liquid.

Ignition switch The main power switch, generally key operated, of a vehicle.

Inertia switch A switch found in the fuel pump circuit to turn off the fuel pump, and other vehicle accessories, in the event of a collision.

Interior lights Lighting in the interior of a vehicle, often called courtesy lights.

Linkage Levers or rods used to transmit power for one part to another.

Load A device connected to an electrical circuit to provide resistance and/or control the current flow.

Load test An electrical test for motors and batteries in which current draw and voltage is measured.

Module a control assembly designed to perform one or more specific tasks.

Ohm A unit of measure of electrical resistance.

Onboard computer The resident or main computer in a vehicle.

Open circuit An incomplete electrical circuit.

Power-steering pump A hydraulic pump used to provide a fluid boost for ease in vehicle steering.

Printed circuit Electrically conductive circuit paths generally etched on a rigid or flexible strata.

Printed circuit board An insulated board on which a printed circuit is etched.

Schematic A drawing of a system using symbols to represent components.

Sending unit An electrical or mechanical sensing device to transmit certain conditions to a remote meter or gauge.

Series A part of an electrical circuit whereby one component is connected to another, as negative to positive, and so on.

Servo A device that converts hydraulic pressure to mechanical movement.

Short circuit The intentional or unintentional grounding of an electrical circuit.

Shunt A parallel electrical connection or circuit.

Speed sensor An electrical device that senses the speed of a rotating shaft or vibrating member.

Squeal A continuous high-pitched noise.

Starter The electric motor and drive used to start a vehicle engine.

Starter solenoid A magnetic switch used to engage the starter for starting an engine.

Switch A mechanical device used to open and close an electrical circuit.

Transmission A device used to couple a motor to a mechanical mechanism.

Troubleshoot To determine the problem, the cause of the problem, and the solution by systematic reasoning.

Turn signal Lights on the four corners of a vehicle to signal a turn.

Wiper A mechanical arm that moves back and forth over the windshield to remove water.

Notes

Notes

Notes

Notes

Notes

Notes

Notes

Notes